Contents

Chapter 1	Problem solving	2
Chapter 2	Algebra	5
Chapter 3	Coordinate geometry	17
Chapter 4	Sequences and series	31
Chapter 5	Functions and transformations	41
Chapter 6	Differentiation	53
Chapter 7	Integration	66
Chapter 8	Trigonometry	76
Formula sheet		93

1 Problem solving

1 The number of eggs in a magic basket doubles every minute. At 11:00 the basket was full.
What time was the basket half full?

2 Every time you cross a magic bridge, the money you have doubles. At the end of each crossing you must pay a toll of \$4. At the end of three crossings, the toll takes all of your money.
How much money did you start with?

3 At a game of rugby, a screen shows the statistics for the percentage of successful shots at goal by a single player, for both the current game and the competition as a whole.

David Burnett, NEW ZEALAND	
Today	**Competition**
3 from 7 attempts	16 from 25 attempts
43%	64%

How many consecutive shots at goal does this player need to succeed with so that his percentage success rate for the game in progress is at least as good as his overall rate for the competition?

4 If $\frac{1}{x+5} = 4$, find the value of $\frac{1}{x+8}$ (without finding the value of x).

QUESTION & WORKBOOK

Cambridge
International AS & A Level

Mathematics

Pure
Mathematics 1

Greg Port

HODDER EDUCATION

Answers to all the questions can be found at www.hoddereducation.com/cambridgeextras

Questions from Cambridge International AS & A Level Mathematics papers are reproduced by permission of Cambridge Assessment International Education. Cambridge Assessment International Education bears no responsibility for the example answers to questions taken from its past question papers which are contained in this publication.

This text has not been through the Cambridge International endorsement process.

Every effort has been made to trace all copyright holders, but if any have been inadvertently overlooked, the Publishers will be pleased to make the necessary arrangements at the first opportunity.

Hachette UK's policy is to use papers that are natural, renewable and recyclable products and made from wood grown in well-managed forests and other controlled sources. The logging and manufacturing processes are expected to conform to the environmental regulations of the country of origin.

Orders: please contact Hachette UK Distribution, Hely Hutchinson Centre, Milton Road, Didcot, Oxfordshire, OX11 7HH. Telephone: +44 (0)1235 827827. Email: education@hachette.co.uk. Lines are open from 9 a.m. to 5 p.m., Monday to Saturday, with a 24-hour message answering service. You can also order through our website at www.hoddereducation.com.

ISBN: 978 1 5104 2184 4

Published by
Hodder Education, an Hachette UK Company
Carmelite House, 50 Victoria Embankment
London EC4Y 0DZ

www.hoddereducation.com

Impression number 10 9 8 7 6 5

Year 2022

Cover photo by Shutterstock/Lodislav Berecz

Illustrations by Integra Software Services

Typeset in Minion Pro Regular 10.5/14 by Integra Software Services Pvt. Ltd., Pondicherry, India

Printed in the UK

A catalogue record for this title is available from the British Library.

5 If p people can do a job in d days, how many days does it take for $p + r$ people to do the same job?

6 A suitcase weighs 24 kg when full and 13 kg when half full.
How much does the suitcase weigh when empty?

7 A fish tank is 100 cm wide and 40 cm high. It is tilted as shown with the water level reaching C, the midpoint of AB. Find the depth of water in the fish tank when it is returned to a horizontal position.

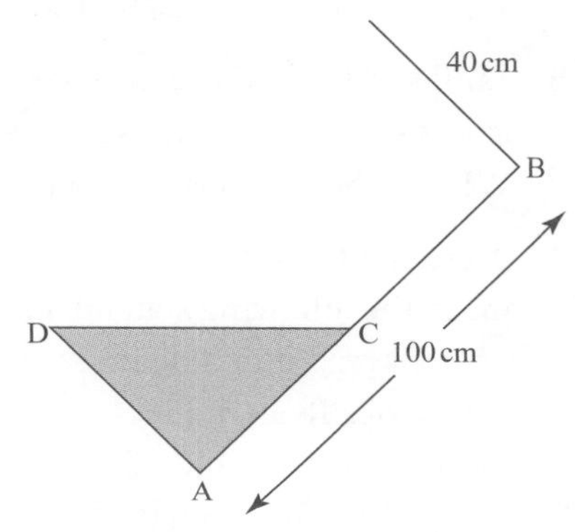

8 There are 20 questions in a multiple-choice test. Five marks are awarded for a correct answer and two marks deducted for an incorrect answer. No marks are awarded for any question left unanswered.
If a student scores 48 marks, what is the *greatest* possible number of questions she answered correctly in the test?

9 Prove that the square of any even number is an even number.

10 Prove that the square of any odd number is odd.

Further practice

1 Two friends are collecting marbles. One says: 'If you give me one of your marbles we will have an equal number of marbles.' The other says: 'If you give me one marble, then I will have double the number you have.'
What is the total number of marbles the friends have?

2 Alma asked his friend Peta about her children. 'I have three children; the product of their ages is 72 and the sum of their ages is your house number,' said Peta. Alma thought for a moment and replied: 'I need more information to find their ages.' Peta replied: 'My oldest daughter is learning to play the piano.' What are the three children's ages?

3 Three men paid \$300 for a suite at a hotel. The manager gave them a discount and asked the lift attendant to give them \$50 back. The lift attendant, confused about how to split the \$50 between the three men, gave them each \$10 back and kept \$20 for himself. So the suite cost the men \$270 in total (\$90 each) and the lift attendant had \$20.
What happened to the other \$10?

4 A plane travels at 350 km/h in still air. There is a wind blowing at a constant speed of 50 km/h. Against the wind, the total time for a journey is 150 minutes.
What will be the time, in minutes, for the return journey with the wind?

5 Sanil's watch loses 10 minutes every hour. He sets his watch correctly at 9 a.m. when he arrives at school and he leaves school when his watch says 5 p.m.
What is the actual time when he leaves school?

6 A group of four boys were accused of a crime that only one of them committed. When asked for their stories, they made the following statements:

Dave: 'I didn't do it.'
Kyle: 'Ryan did it.'
Ryan: 'Dave did it.'
Leon: 'Ryan is lying.'

If only one of the four suspects is telling the truth, who is guilty?

7 A hot tap can fill a bath in 12 minutes and the cold tap can fill the bath in 6 minutes.
How long does it take to fill the bath with both taps on?

8 I walk at 4 km/h and run at 6 km/h. I find that I can save $3\frac{3}{4}$ minutes by running instead of walking from my house to the train station.
What is the distance from my house to the train station?

9 A cask is filled with 45 litres of juice. Nine litres are removed and the cask is refilled with water. Then nine litres are removed again and the cask is refilled again with water. What is the ratio of water to juice in the final mixture?

10 a and b are two odd integers such that $a > b$. Prove that $a^2 - b^2$ is a multiple of 8.

2 Algebra

2.1 Background algebra

1 Simplify these expressions as fully as possible.

(i) $2(a-3b)-3(b-3a)$

(ii) $7cd\,(d^2-2)-3cd^2\,(8d+5c^3)$

(iii) $\frac{8f^4}{3g}\times\frac{9g^3}{12fg}$

(iv) $\frac{x-1}{4}+\frac{5-x}{5}$

2 Factorise these expressions fully.

(i) $12mn^2+9mn^3$

(ii) p^2-p-12

3 Solve the following equations.

(i) $2(x+5)=x-7$

(ii) $\frac{1}{2}(6x+8)-3=9-\frac{3}{2}(4-10x)$

4 Hakim drives from Auckland to Hamilton in 2 hours. Ravi leaves at the same time as Hakim and drives the same route at, on average, 4 km/h slower and arrives 6 minutes after Hakim. Find the distance from Auckland to Hamilton.

5 Make the letter in brackets the subject of the formula.

(i) $\frac{v}{b} - c = \frac{d}{e}$ (e)

(ii) $km^2 + n = p - wk$ (k)

6 Simplify these surds.

(i) $\sqrt{18}$

(ii) $\sqrt{75}$

(iii) $\sqrt{1\frac{7}{9}}$

7 Rationalise the denominator in each of the following.

(i) $\frac{2}{\sqrt{5}}$

(ii) $\frac{3}{\sqrt{7} - 1}$

(iii) $\frac{1 + \sqrt{3}}{1 - \sqrt{3}}$

8 Simplify these expressions.

(i) $(\sqrt{2} + 1)(\sqrt{2} - 1)$

(ii) $(1 - \sqrt{b})(2 + \sqrt{b})$

2.2 Quadratic equations

1 Solve these quadratic equations by factorising.

(i) $x^2 + 5x = 0$

(ii) $x^2 - 2x - 8 = 0$

(iii) $2x^2 - x - 3 = 0$

(iv) $3x^2 - 6x = 0$

(v) $x^2 - 3x - 40 = 0$

(vi) $6x^2 + 7x - 3 = 0$

2 Solve the following equations.

(i) $x^4 + 3x^2 - 4 = 0$

(ii) $5 - \frac{2}{x} = 2x$

(iii) $x + 2\sqrt{x} = 8$

(iv) $x^6 + 8 = 9x^3$

2.3 Completing the square

1 Write these quadratic expressions in the completed square form $(x \pm a)^2 \pm b$.

(i) $x^2 - 6x + 1$

(ii) $x^2 + 4x$

2 Using your answers to question 1, solve the following equations.

(i) $x^2 - 6x + 1 = 0$

(ii) $x^2 + 4x = 0$

3 Write these quadratic expressions in the form $a(x \pm b)^2 \pm c$.

(i) $2x^2 - 4x + 7$

(ii) $3x^2 + 12x - 4$

(iii) $4x^2 + 24x - 16$

(iv) $9x^2 - 6x$

4 Write $3 - 8x - x^2$ in the form $a - (x + b)^2$, stating the values of a and b.

2.4 The graphs of quadratic functions

1 Sketch these quadratic curves, and state the coordinates of the vertex.

(i) $y = (x + 1)^2$

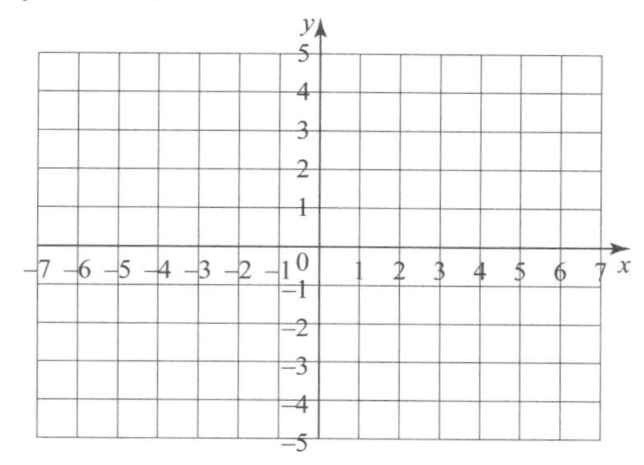

Vertex: (——, ——)

(ii) $y = (x + 4)^2 - 2$

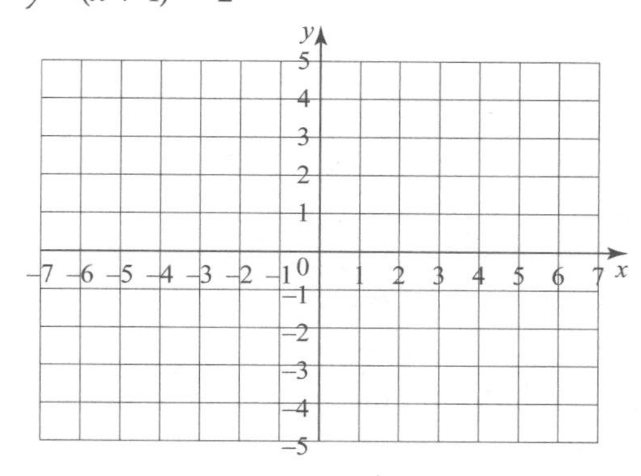

Vertex: (——, ——)

(iii) $y = -(x - 2)^2 + 1$

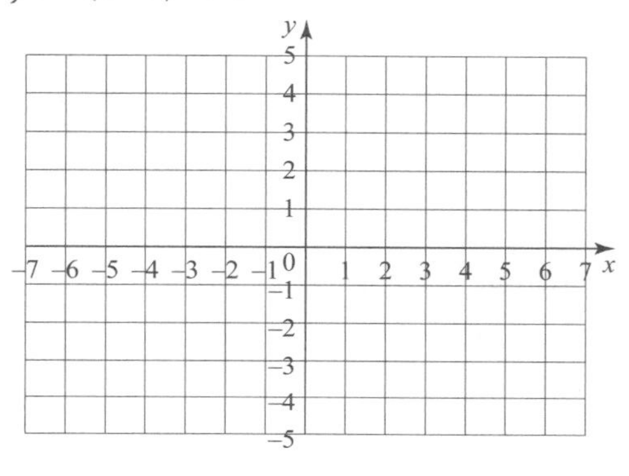

Vertex: (——, ——)

(iv) $y = 2(1 - 2x)^2 - 4$

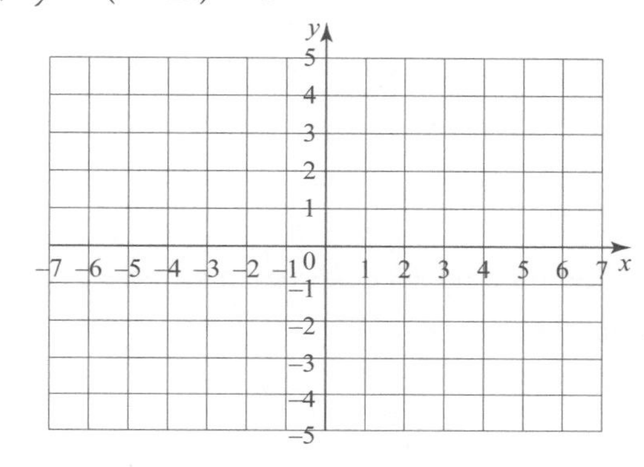

Vertex: (——, ——)

2 Write the equation of these graphs in the form $y = (x + b)^2 + c$. (The coefficient of the x^2 term is 1.)

(i)

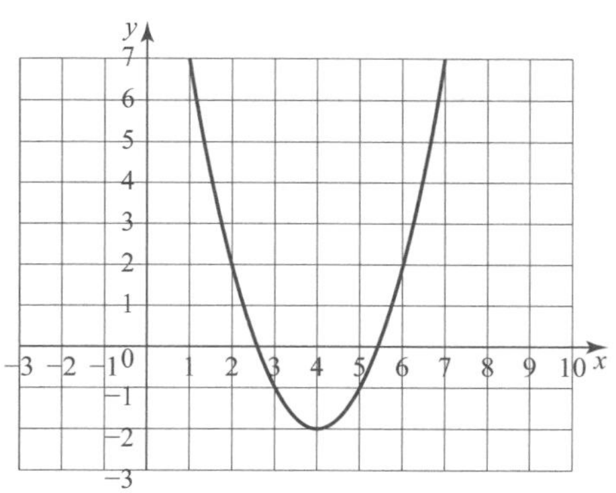

____

(ii)

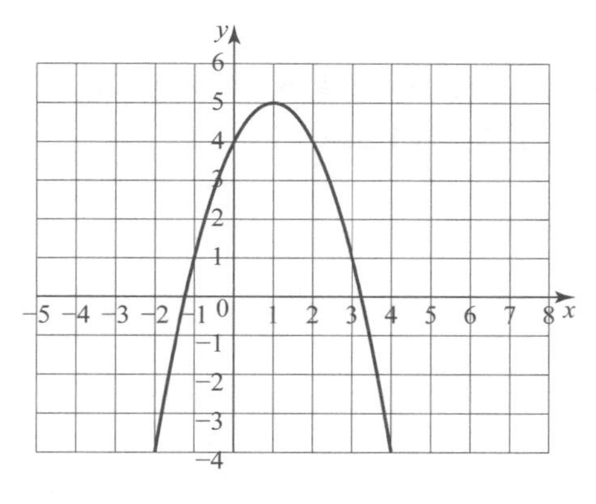

____

2.5 The quadratic formula

1 Solve the following equations using the quadratic formula.

(i) $x^2 + x - 5 = 0$

(ii) $2x^2 - 5x = 6$

2 Find the value of the discriminant for these quadratic equations, and hence state the number of real solutions for each equation.

(i) $x^2 - 2x + 4 = 0$

(ii) $4x^2 + 4x + 1 = 0$

3 (i) Find the value(s) of k in each of these equations so that the equation has **one** real solution.

(a) $kx^2 + 4x - 1 = 0$

(b) $x^2 + kx + k - 1 = 0$

(ii) Find the value(s) of k in each of these equations so that the equation has **two** real solutions.

(a) $x^2 - 2x + k = 0$

(b) $kx^2 - kx + 1 = 0$

(iii) Find the value(s) of k in each of these equations so that the equation has **no** real solutions.

(a) $2x^2 - 2kx + 1 = 0$

(b) $kx^2 - 4x + 2k = 0$

4 The quadratic equation $x^2 + mx + n = 0$, where m and n are constants, has roots 6 and -2.

(i) Find the values of m and n.

(ii) Using these values of m and n, find the value of the constant p such that the equation $x^2 + mx + n = p$ has equal roots.

2.6 Simultaneous equations

1 Solve the equations simultaneously. Both equations are linear equations.

(i) $x - y = 4$
$x + 2y = 1$

(ii) $2x + 3y = 11$
$3x + y = -1$

2 Solve the equations simultaneously.

(i) $y = x^2 + 4x$
$y = x - 2$

(ii) $y = x^2 + 4x - 2$
$2x + y = -2$

(iii) $x + y + 1 = 0$
$x^2 + y^2 = 25$

(iv) $xy = 3$
$2x + y = 7$

2.7 Inequalities

1 Solve these inequalities.

(i) $3(x+4) \leqslant -15$ **(ii)** $1 - \frac{3x}{4} > 7$ **(iii)** $-4x - 1 < -x + 5$

2 Solve these quadratic inequalities.

(i) $x(x-1) > 0$ **(ii)** $2x(1-x) \geqslant 0$

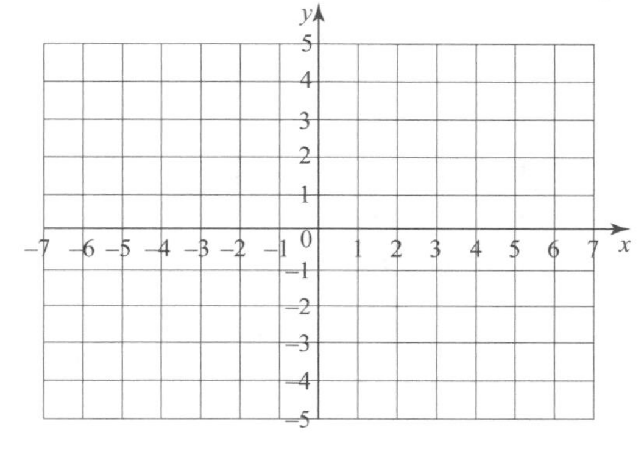

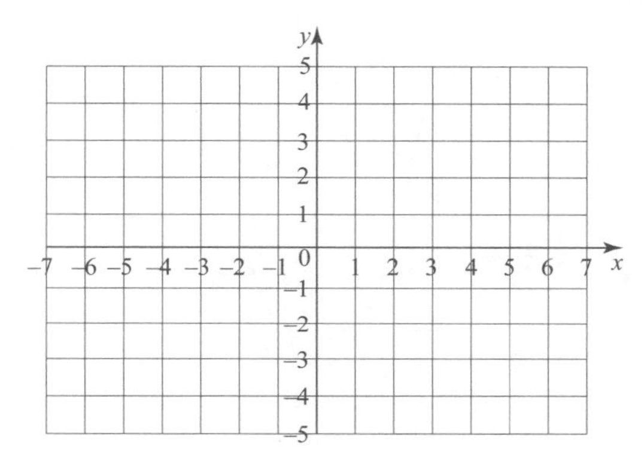

(iii) $(x+1)(x-1) > 0$ **(iv)** $x^2 < 9$

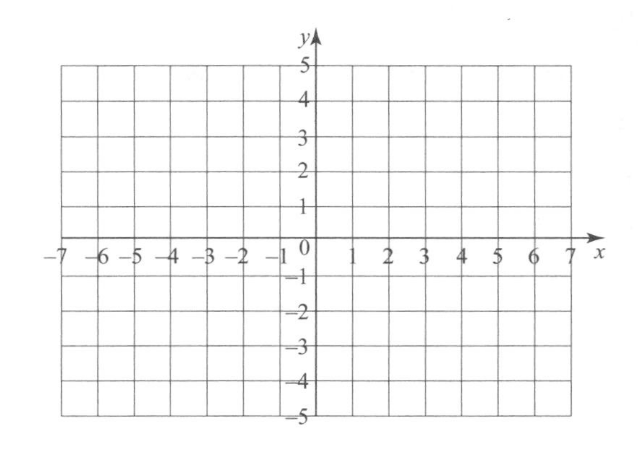

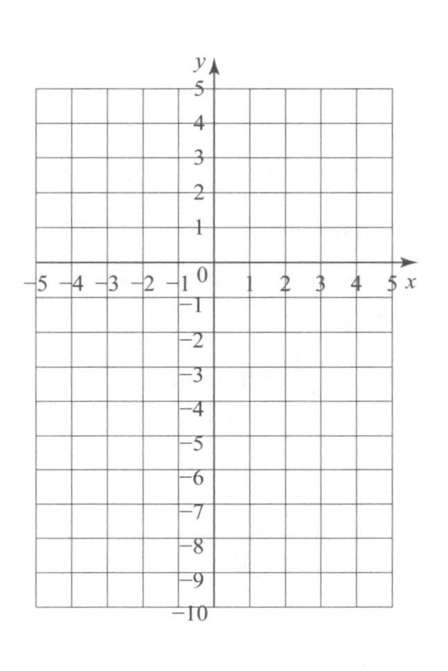

Further practice

1 Factorise fully.

(i) $3q^2 + 5q - 2$ **(ii)** $ts + tp - 2us - 2up$

2 Make the letter in brackets the subject of the formula.

(i) $\frac{1}{f} = \frac{1}{u} + \frac{1}{v}$ (v) **(ii)** $\sqrt{d - 3e} = \frac{1}{2\pi}\sqrt{\frac{p}{w}}$ (e)

3 Solve these equations.

(i) $x^2 - 25 = 0$ **(ii)** $x^2 + 5x - 14 = 0$

(iii) $3x^2 - 5x - 2 = 0$ **(iv)** $5x^2 + 13x - 6 = 0$

(v) $\frac{3}{x^4} - \frac{11}{x^2} = 4$ **(vi)** $\frac{2}{x} - 1 = 4 - \frac{9}{\sqrt{x}}$

4 Write these quadratic expressions in the form $a(x \pm b)^2 \pm c$.

(i) $2x^2 + 12x + 11$ **(ii)** $5x^2 - 40x + 72$

5 Write the expression $7 + 8x - 2x^2$ in the form $a - b(x - c)^2$, stating the values of a, b and c.

6 Write the equation of these graphs in the form $y = (x + b)^2 + c$. (The coefficient of the x^2 term is 1.)

(i)

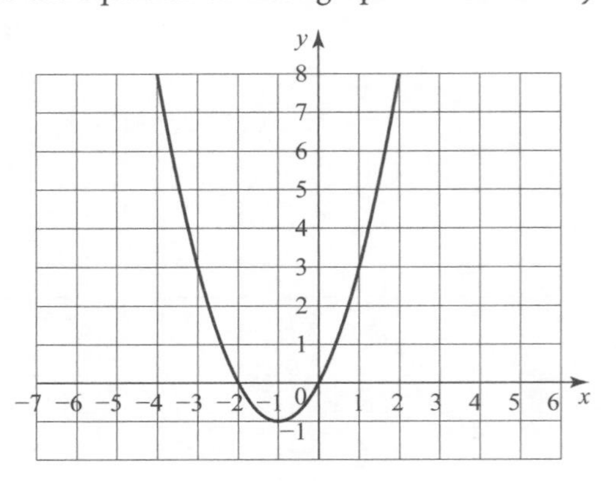

(ii)

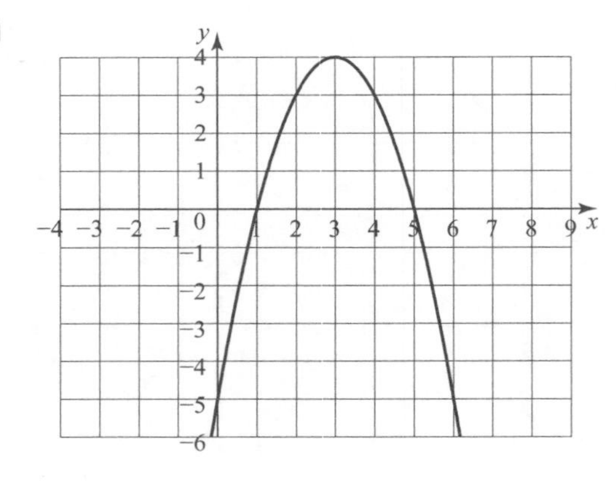

7 Solve these equations using the quadratic formula.

(i) $1 - 3x^2 = 5x$ **(ii)** $12x = 6x^2 - 5$

8 If $\sqrt{90} = a\sqrt{10}$ find the value of a.

9 Find the value of the discriminant for these quadratic equations, and hence state the number of real solutions for each equation.

(i) $-x^2 - 3x - 2 = 0$ **(ii)** $x^2 + 4 = 0$ **(iii)** $2 - x + 5x^2 = 0$ **(iv)** $-4x^2 + 3x = 0$

10 (i) Find the value(s) of k in each of these equations so that the equation has **one** real solution.

(a) $4 + kx + x^2 = 0$ **(b)** $kx^2 = kx + 1$

(ii) Find the value(s) of k in each of these equations so that the equation has **two** real solutions.

(a) $3x^2 - kx + 3 = 0$ **(b)** $3 - 2kx^2 = 6x$

(iii) Find the value(s) of k in each of these equations so that the equation has **no** real solutions.

(a) $k - x + 9x^2 = 0$ **(b)** $2x + kx^2 = 1$

11 Solve these equations simultaneously. Both equations are linear.

(i) $4x - \frac{1}{2}y = 6$

$3x - 2y = -15$

(ii) $2x + 3y = 3$

$y = 9 - 2x$

12 Solve these equations simultaneously.

(i) $y = x^2 + 9$
$y - 4x = 5$

(ii) $(x - 2)^2 + y^2 = 5$
$2x + y = 9$

(iii) $x^2 + 2y^2 = 9$
$x + y = 1$

(iv) $2x + 3y = 7$
$3x^2 = 4 + 4xy$

13 Solve these quadratic inequalities.

(i) $16 - x^2 \geqslant 0$

(ii) $x^2 + 5x > 0$

(iii) $x^2 - 2x - 8 \leqslant 0$

(iv) $x^2 + 5x - 14 > 0$

(v) $(5 + x)(3 - 2x) < 0$

(vi) $2x^2 - 5x - 3 < 0$

(vii) $3 - 8x - 3x^2 \geqslant 0$

(viii) $3x^2 < 6x$

14 (i) Write the expression $4x^2 + 32x + 70$ in the form $a(x + b)^2 + c$ and hence state the coordinates of the vertex of the graph of $y = 4x^2 + 32x + 70$.

(ii) Find the values of x when $y < 22$.

Past exam questions

1 Solve the inequality $x^2 - x - 2 > 0$. [3]

Cambridge International AS & A Level Mathematics 9709 Paper 13 Q1 November 2013

2 (i) Express $x^2 + 6x + 2$ in the form $(x + a)^2 + b$, where a and b are constants. [2]

(ii) Hence, or otherwise, find the set of values of x for which $x^2 + 6x + 2 > 9$. [2]

Cambridge International AS & A Level Mathematics 9709 Paper 11 Q1 November 2016

3 (i) Express $2x^2 - 10x + 8$ in the form $a(x + b)^2 + c$, where a, b and c are constants, and use your answer to state the minimum value of $2x^2 - 10x + 8$. [4]

(ii) Find the set of values of k for which the equation $2x^2 - 10x + 8 = kx$ has no real roots. [4]

Cambridge International AS & A Level Mathematics 9709 Paper 13 Q8 June 2014

STRETCH AND CHALLENGE

1 **(i)** For a quadratic equation $ax^2 + bx + c = 0$ with roots α and β, show that

$$\alpha + \beta = -\frac{b}{a} \text{ and } \alpha\beta = \frac{c}{a}$$

(ii) Using this fact, find the values of k for the equation $4x^2 + (k + 2)x + 72 = 0$ such that one root is double the other.

(iii) The roots of the equation $3x^2 - 4x + 7 = 0$ are α and β. Find the quadratic equation with roots $\frac{1}{\alpha}$ and $\frac{1}{\beta}$.

(iv) If α and β are the roots of the equation $x^2 - 2x + 3 = 0$, find the quadratic equation whose roots are α^3 and β^3.

2 Solve $9^x - 3^{x+1} - 54 = 0$.

3 Find all the possible values of k such that the equation $k2^x + 2^{-x} = 8$ has a single root. Find the root in this case.

3 Coordinate geometry

3.1 The length, gradient and midpoint of a line

1 Find the gradient of the following straight lines.

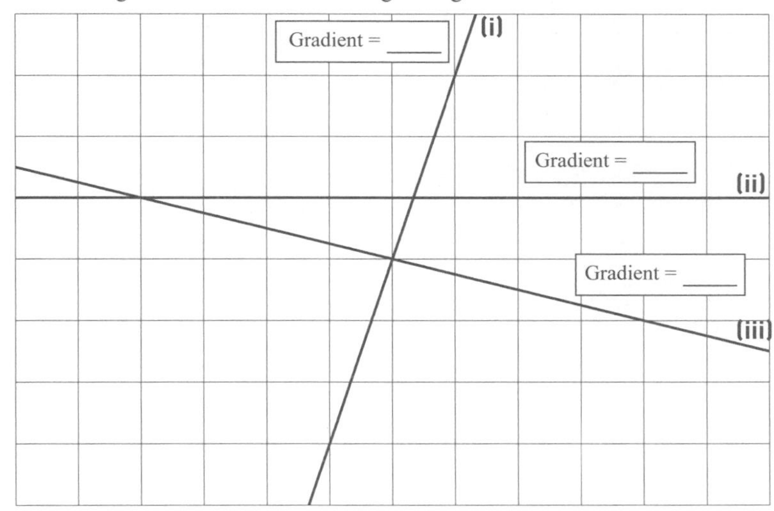

2 Given the coordinates of the end points of these lines, find the length, midpoint and gradient of each line.

(i) A(3, 1) and B(–1, –1)

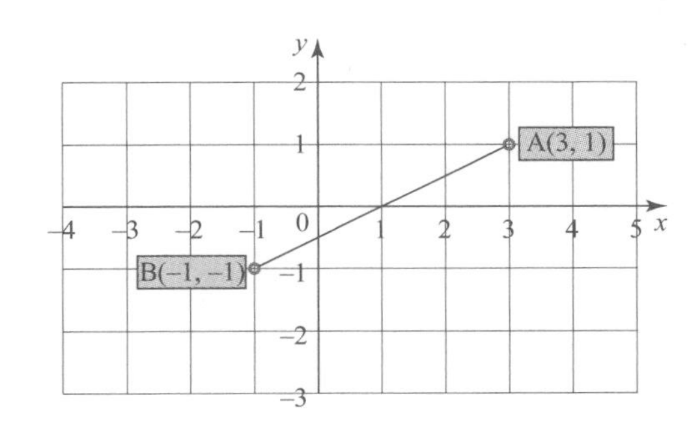

Length: ________

Midpoint: ________

Gradient: ________

(ii) C(12, –3) and D(8, 7)

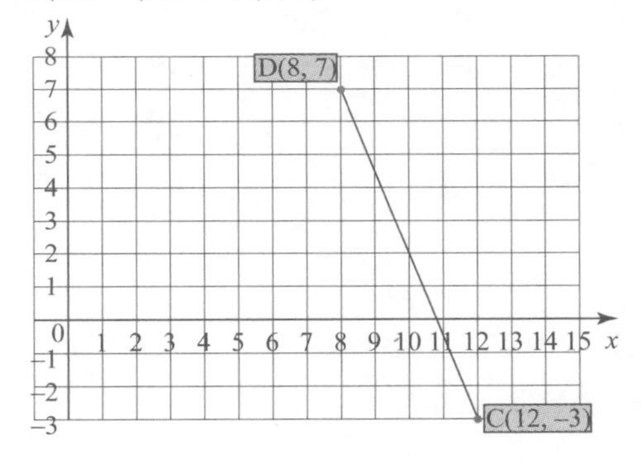

Length: ________

Midpoint: ________

Gradient: ________

3 The midpoint of the line joining A(–1, 5) and B(m, n) is (2, 5). Find the values of m and n.

4 C is the point (–2, 1) and D is the point (x, 3).

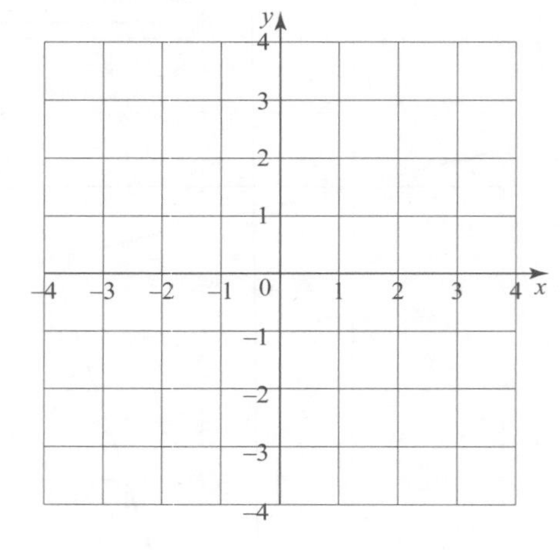

Find the value of x if the gradient of the line CD is:

(i) 1

(ii) –1

(iii) $\frac{2}{3}$

5 (i) Find the angle that each of these lines makes with the positive x-axis.

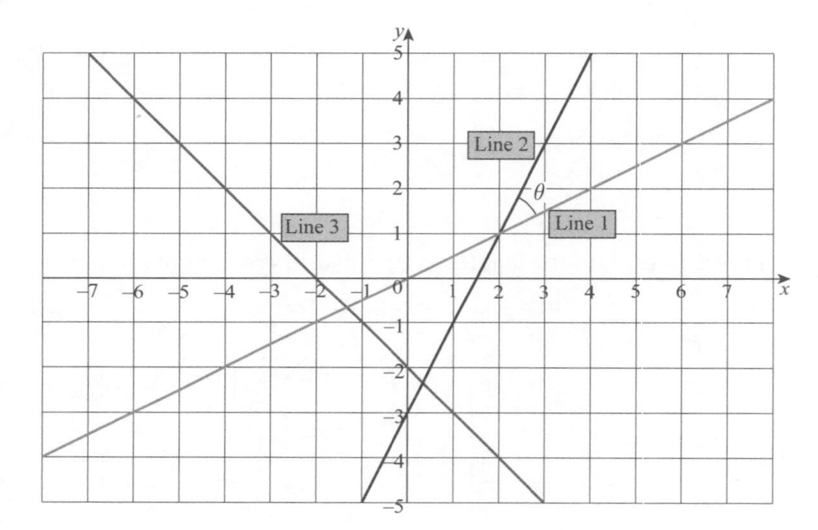

Line 1: ________ Line 2: ________ Line 3: ________

(ii) Find θ, the angle between Line 1 and Line 2.

3.2 The equation of a straight line

1 Write down the gradient and y-intercept of the following lines. Sketch the lines on the grid below.

(i) $y = 3x - 1$

Gradient = ________

y-intercept = ________

(ii) $y = -2x + 3$

Gradient = ________

y-intercept = ________

(iii) $y = \frac{3x}{2} + 1$

Gradient = ________

y-intercept = ________

(iv) $y = -\frac{x}{2} - 2$

Gradient = ________

y-intercept = ________

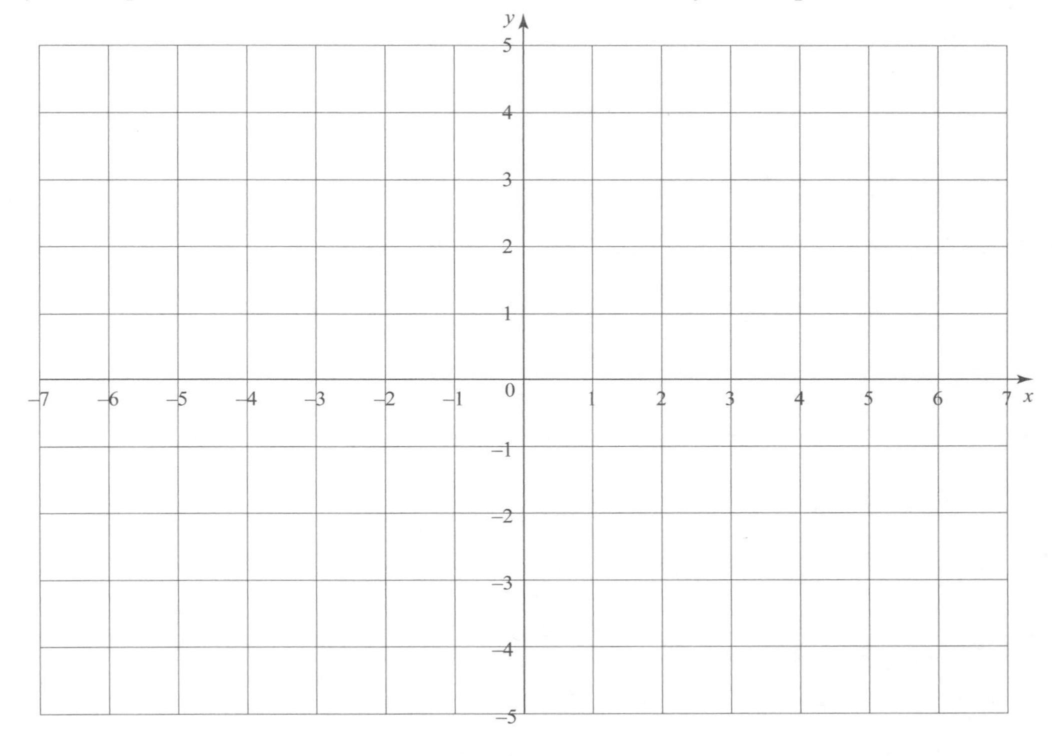

2 For each equation below:

- rearrange the equation so it is in the form $y = mx + c$
- hence state the gradient and y-intercept
- draw the line on the grid below.

(i) $x + y = 2$

Gradient = ________

y-intercept = ________

(ii) $3x - y = 2$

Gradient = ________

y-intercept = ________

(iii) $2x + 4y - 9 = 0$

Gradient = ________

y-intercept = ________

(iv) $3x - 2y = 8$

Gradient = ________

y-intercept = ________

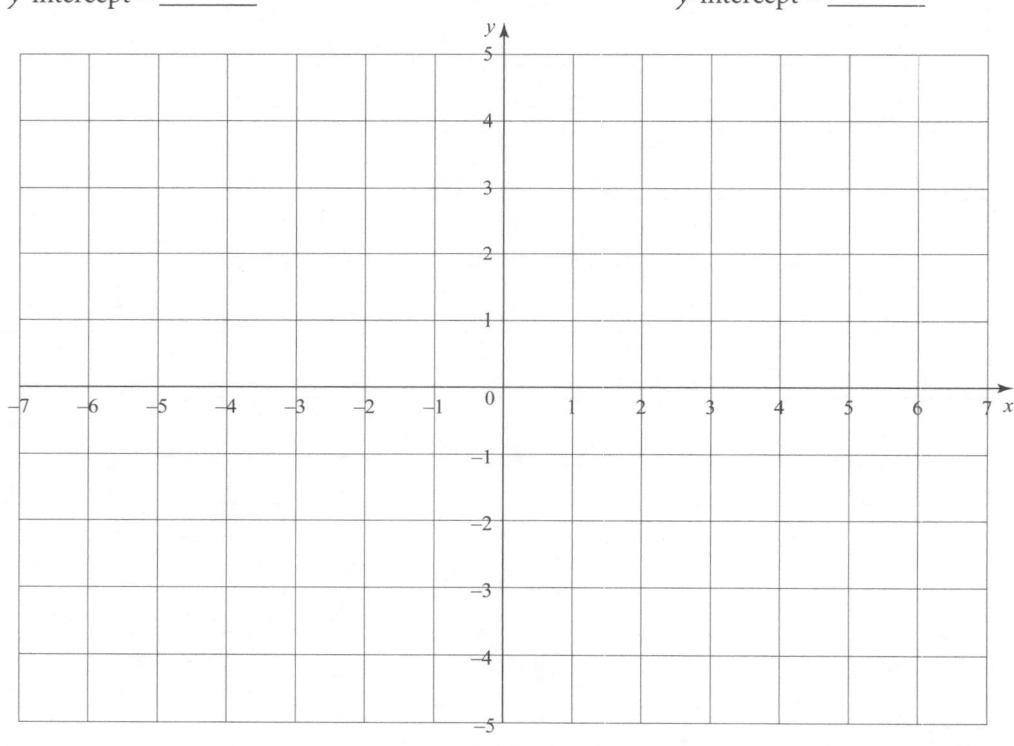

3 Rearrange the equations of these lines so that they are in the form $ax + by + c = 0$.

(i) $y = -\frac{x}{3} - 2$

(ii) $y = \frac{4x}{5} + \frac{1}{3}$

4 Write down the equations of the lines drawn on the grid below.

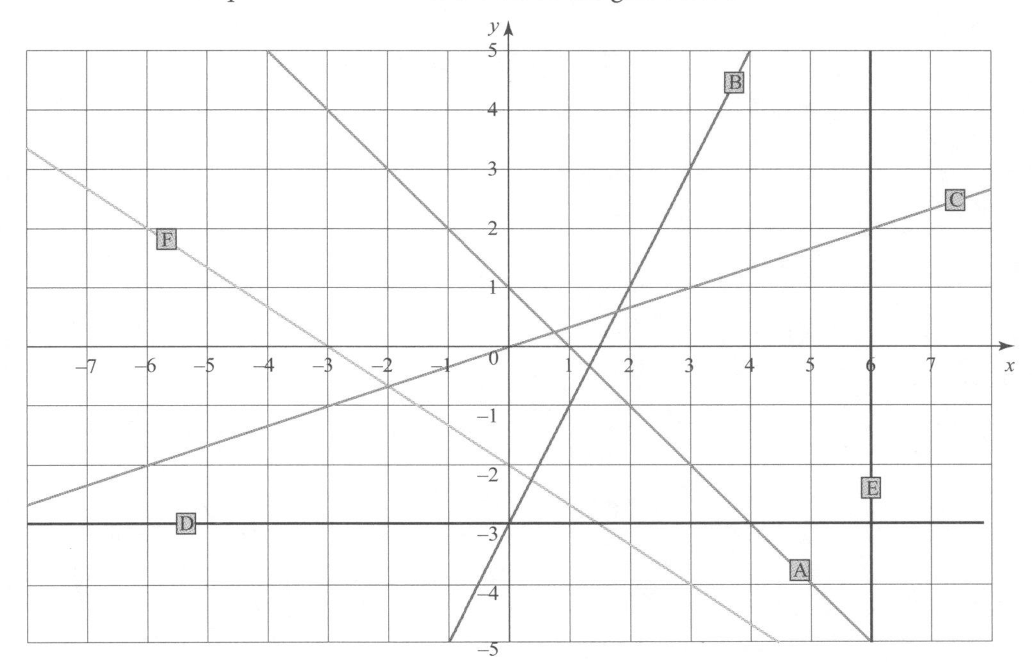

A: ______________________ B: ______________________

C: ______________________ D: ______________________

E: ______________________ F: ______________________

5 Fill in the spaces in the table.

Gradient of line	Gradient of perpendicular
1	
	-4
$-2\frac{1}{3}$	
	$\frac{2}{3}$
0.3	

6 Find the equation of the line that is:

(i) parallel to $2x - y = 1$ going through (4, 1)

(ii) perpendicular to $2x - y = 1$ going through (−3, 1).

(iii) Draw both lines on the grid.

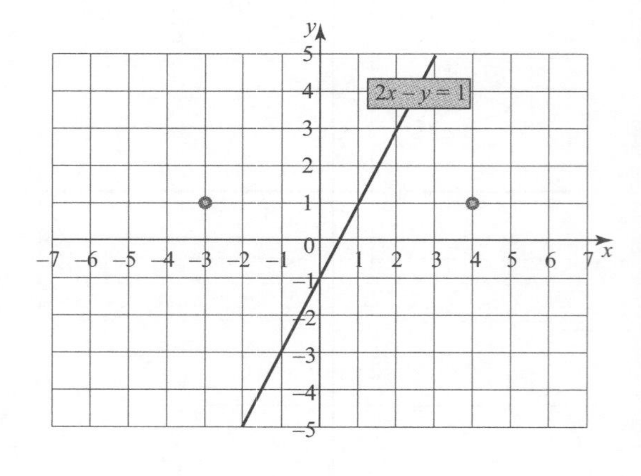

3.3 The intersection of two lines

1 Find the area of the triangle formed by the straight line $3x + 2y = 8$ and the x- and y-axes.

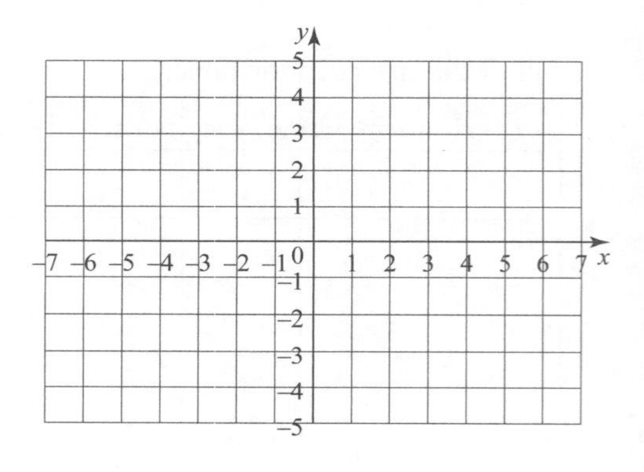

2 Find the equation of the perpendicular bisector of the line joining A(2, 4) and B(−6, 0).

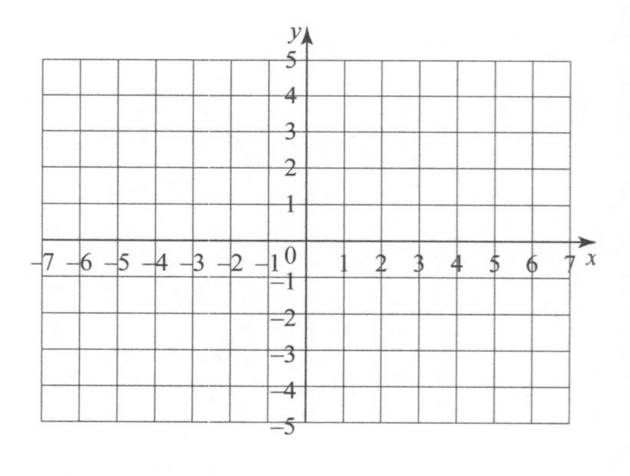

3 A *median* of a triangle is a line that joins a vertex to the midpoint of the opposite side.

A triangle is formed by the points A(–5, 2), B(2, 3) and C(4, –5).

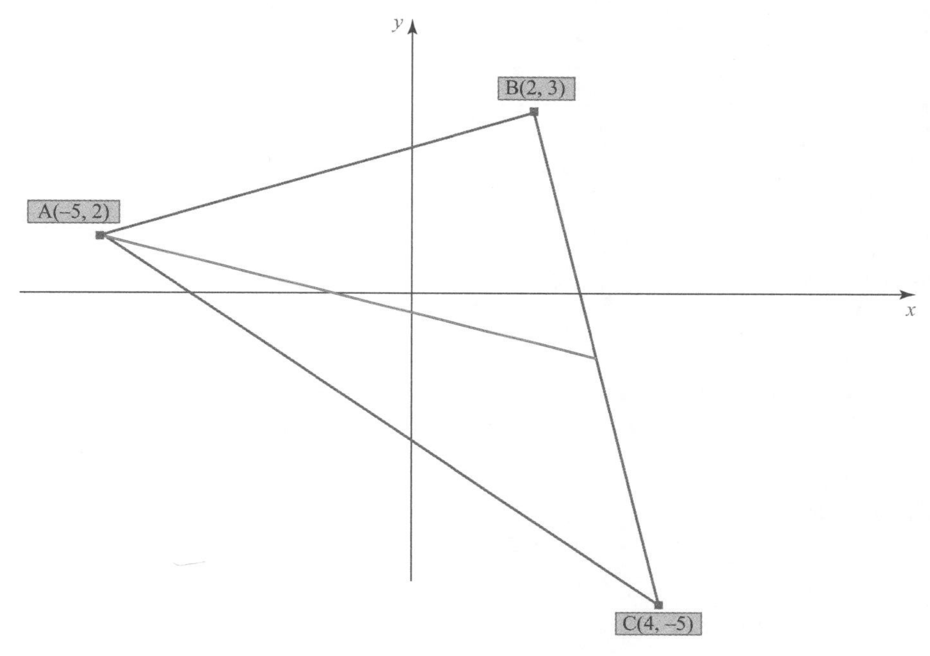

Find the equation of the median from the point A, as shown.

4 Given that $ax - 3y + 1 = 0$ and $2x + by - 6 = 0$ are perpendicular lines, find the ratio $a:b$.

5 The line $y = 9 - 2x$ is a tangent to the curve $y = x(4 - x)$ at the point where $x = 3$, as shown. Find the area of the triangle formed by the tangent, the normal and the x-axis.

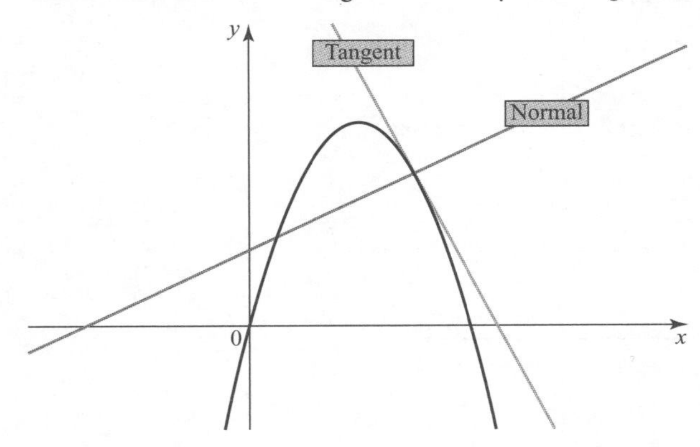

3.4 The circle

1 Find the centre and radius of these circles.

(i) $(x-1)^2 + (y+3)^2 = 16$

(ii) $x^2 + y^2 + 6x - 2y + 1 = 0$

2 Find the equations of the circles with the following properties:

(i) radius 6 and centre (4, –1)

(ii) radius $\sqrt{8}$ and centre (–2, 7)

(iii) centre (2, 4) and passing through (–1, 0)

(iv) centre (3, –3) and touching both the x- and y-axes.

3 Find the equations of the two circles shown.

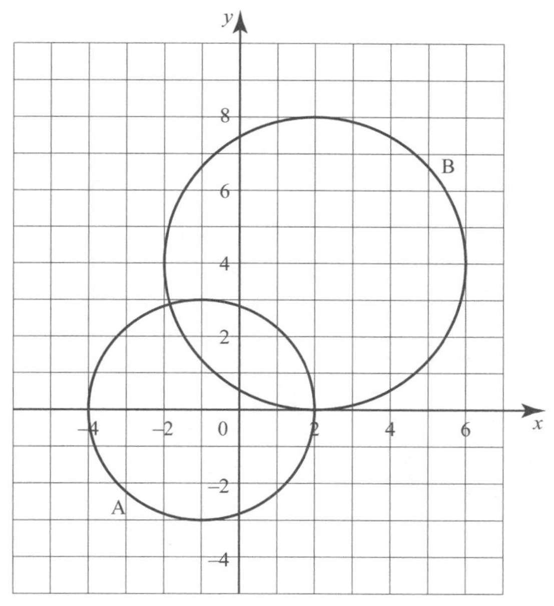

A: ______________________

B: ______________________

4 The circle $x^2 + y^2 - 2x + 4y - 20 = 0$ has centre C and passes through points A and B.

(i) State the coordinates of C.

It is given that the midpoint, D, of AB has coordinates $(2.5, 2.5)$.

(ii) Find the equation of AB, giving your answer in the form $y = mx + c$.

(iii) Find, by calculation, the x-coordinates of A and B.

3.5 The intersection of a line and a curve

1 Find the coordinates of the point(s) of intersection.

(i) $y = x^2 + 4x$

$y = x - 2$

(ii) $y = x^2 + 4x - 2$

$2x + y = -2$

(iii) $xy = 3$

$2x + y = 7$

(iv) $(x - 2)^2 + y^2 = 5$

$2x + y = 9$

2 Find the value of k such that the line $y - x = k$ is a tangent to the curve $y = x^2 + 3x + 9$.

3 The curve $y = 2 - x^2$ and the line $kx + y = 3$ intersect at two points.
Find the possible values of k.

4 A circle has equation $(x+1)^2 + y^2 = 25$. The tangent to the circle at $(-4, 4)$ is $y = mx + c$.

(i) Find the values of m and c.

(ii) The tangent intersects the x-axis at the point A and the y-axis at the point B.
Find the area of the triangle AOB.

5 Find the value(s) of the constant k for which the line $y + kx = 8$ is a tangent to the curve $y = 2x^2 + 3x + 10$.

6 A curve has equation $y = 2 + kx^2$ and a line has equation $y = kx + 1$, where k is a non-zero constant.

(i) Find the set of values of k for which the curve and the line have no common points.

(ii) State the value of k for which the line is a tangent to the curve and, for this case, find the coordinates of the point where the line touches the curve.

Further practice

1 Given the coordinates of the end points of these lines, find the length, midpoint and gradient of each line.

(i) E(–5, 3) and F(3, –1)   **(ii)** G(–4, 3) and H(–10, –9)

2 Line A has a gradient of 2 and line B has a gradient of $-\frac{1}{3}$. Find the acute angle between the lines.

3 Sketch the following lines and state their gradient and y-intercept.

(i) $y = \frac{1}{4}x - 3$ **(ii)** $y = 5 - \frac{2x}{3}$

4 Rearrange the equation of each line so it is in the form $y = mx + c$.

(i) $x - 2y + 1 = 0$ **(ii)** $5x + 10y + 4 = 0$

5 Rearrange the equation of each line so it is in the form $ax + by + c = 0$.

(i) $y = \frac{3}{2}x + 4$ **(ii)** $y = -\frac{8}{5}x - 1$

6 Find the equation of the line

(i) parallel to $4x + 3y = 1$ going through (–2, 0)   **(ii)** perpendicular to $4x + 3y = 1$ going through (3, –1).

Give your answers in the form $ax + by + c = 0$, where a, b and c are integers.

7 Points A, B and C have coordinates (4, 1), (6, –2) and (–1, –9) respectively.

(i) Find the coordinates of the midpoint of AC.

(ii) Find the equation of the line through B perpendicular to AC.
Give your answer in the form $ax + by + c = 0$.

8 The line L_1 is given by $x - 3y = 6$. The point P is (–3, 2).

(i) Find the equation of the line perpendicular to L_1 that goes through P.

(ii) Find the perpendicular distance from the point P to L_1.

9 Find the centre and radius of these circles.

(i) $x^2 + (y + 5)^2 = 64$ **(ii)** $x^2 + y^2 + 4x - 2y - 4 = 0$

10 Find the equations of the circles with the following properties:

(i) centre (5, –1) and radius 8

(ii) radius 3 and with the positive x-axis and negative y-axis as tangents.

11 Find the coordinates of the point(s) of intersection between each curve and straight line.

(i) $y = x^2 + 9$
$y - 4x = 5$

(ii) $x^2 + 2y^2 = 9$
$x + y = 1$

12 The curve $y = 2 - x^2$ and the line $kx + y = 3$ intersect at two points. Find the possible values of k.

13 Find the values of k so the curve $xy = k$ has no points of intersection with the line $2x + y = 3$.

14 Find the value(s) of k that make the line $y - kx = -\sqrt{8}$ a tangent to the curve $x^2 - xy = k$.

Past exam questions

1 The point A has coordinates (–2, 6). The equation of the perpendicular bisector of the line AB is $2y = 3x + 5$.

(i) Find the equation of AB. [3]

(ii) Find the coordinates of B. [3]

Cambridge International AS & A Level Mathematics 9709 Paper 12 Q2 June 2017

2 Two points have coordinates A(5, 7) and B(9, –1).

(i) Find the equation of the perpendicular bisector of AB. [3]

The line through C(1, 2) parallel to AB meets the perpendicular bisector of AB at the point X.

(ii) Find, by calculation, the distance BX. [5]

Cambridge International AS & A Level Mathematics 9709 Paper 12 Q5 March 2016

3 The point A has coordinates $(p, 1)$ and the point B has coordinates $(9, 3p + 1)$, where p is a constant.

(i) For the case where the distance AB is 13 units, find the possible values of p. [3]

(ii) For the case in which the line with equation $2x + 3y = 9$ is perpendicular to AB, find the value of p. [4]

Cambridge International AS & A Level Mathematics 9709 Paper 13 Q7 June 2015

4 A line has equation $y = 2x - 7$ and a curve has equation $y = x^2 - 4x + c$, where c is a constant. Find the set of possible values of c for which the line does not intersect the curve. [3]

Cambridge International AS & A Level Mathematics 9709 Paper 13 Q1 November 2015

5 The diagram shows a parallelogram ABCD, in which the equation of AB is $y = 3x$ and the equation of AD is $4y = x + 11$. The diagonals AC and BD meet at the point $E\left(6\frac{1}{2}, 8\frac{1}{2}\right)$.
Find, by calculation, the coordinates of A, B, C and D. [9]

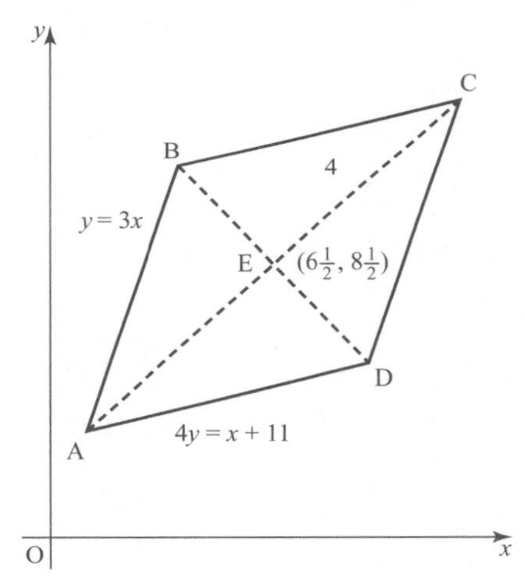

Cambridge International AS & A Level Mathematics 9709 Paper 13 Q11 June 2014

6 Find the set of values of m for which the line $y = mx + 4$ intersects the curve $y = 3x^2 - 4x + 7$ at two distinct points. [5]

Cambridge International AS & A Level Mathematics 9709 Paper 13 Q2 June 2011

STRETCH AND CHALLENGE

1 The line $4x + y + 2 = 0$ is a tangent to the curve $y = 2x^2$ at the point A. The normal to the curve at the same point also goes through the curve at the point B.
Find the length of AB.

2 The line $\frac{x}{a} - \frac{y}{a} = 4$, where a and b are positive constants, meets the x-axis at M and the y-axis at N.
Given that the gradient of the line is $\frac{1}{2}$ and that the length of MN is $\sqrt{720}$, find the values of a and b.

3 The perpendicular bisector of the line joining $A(-6, 1)$ and $B(k, -3)$ passes through the y-axis at -9.
Find the possible value(s) of the constant k.

4 **(i)** For what values of k does the line $y = kx + 1$ intersect the curve $y = x^2 - 4x + 3$ on the x-axis?

(ii) Find an expression for k in terms of a and b where the line $y = kx + 1$ intersects the curve $y = (x - a)(x - b)$ on the x-axis.

(iii) Find an expression for k in terms of a, b and c where the line $y = kx + c$ intersects the curve $y = (x - a)(x - b)$ on the x-axis.

5 For what values of k does the line $kx - 2y = -1$ intersect the curve $y = 2x^2 + x + 1$ at two points?

4 Sequences and series

4.1 Arithmetic progressions

1 Find the required term in each of these arithmetic sequences.

(i) 5, 11, 17, 23, … (12th term)

(ii) −3, −8, −13, … (15th term)

(iii) $\frac{2}{3}, \frac{6}{5}, \frac{26}{15}, \ldots$ (7th term)

(iv) 0.85, 0.55, 0.25, … (11th term)

2 Find the sum of the following arithmetic progressions.

(i) 21, 15, 9, … (9 terms)

(ii) −8, 0, 8, … (18 terms)

(iii) 7, 9, 11, … , 79

(iv) 0.05, 0.15, 0.25, … , 4.05

3 In an arithmetic sequence, the fifth term is 34 and the tenth term is 43.
Find the first term and the sum of the first 20 terms of the sequence.

4 The first three terms of an arithmetic sequence are m, $2m + n$ and $3m + 2n$.
Find the common difference and the 10th term in terms of m and n.

5 Paul buys a car on hire purchase and agrees to pay back the \$3375 with weekly payments that are an arithmetic progression. His first payment is \$40 and the debt is paid in 30 weeks. Find the fifth payment.

6 The seventh term of an arithmetic progression is 32 and the sum of the first five terms is 130.

(i) Find the first term of the progression and the common difference.

The nth term of the progression is 56.

(ii) Find the value of n.

7 The first term of an arithmetic progression is 4 times the value of the fourth term. The sixth term of the progression is 4 less than the fourth term.
Find the value of the eighth term.

4.2 Geometric progressions

1 Find the required term in each of the following geometric sequences.

(i) $\frac{1}{2}$, 1, 2, 4, ... (10th term)

(ii) −16, 8, −4, 2, ... (8th term)

(iii) 0.1, 0.01, 0.001, ... (9th term)

(iv) 6, 9, 13.5, ... (12th term)

2 Find the sum of the following geometric progressions.

(i) 3, 6, 12, 24, ... (8 terms)

(ii) 81, 27, 9, 3, ... (20 terms)

(iii) 2, −5, 12.5, −31.25, ... (14 terms)

(iv) −240, 48, −9.6, 1.92, ... (10 terms)

3 Which of the sequences in question 2 have a sum to infinity? For those sequences, calculate the sum to infinity.

4 Three consecutive terms of a geometric sequence are 32, x, 2. Find the value of x.

5 Find the fifth term of the geometric sequence with a first term of 8 and a sum to infinity of 12.

6 For what values of x does the sequence 3, $6x$, $12x^2$, … have a sum to infinity?

7 The first, third and fifth terms of a geometric sequence are $\frac{5x+1}{2}$, $x + 2$ and $\frac{x}{2}$ respectively.

(i) Find the value of x.

(ii) Given that the common ratio of this geometric sequence is positive, calculate the sum to infinity of the series.

8 Penny decides to start running every day. She has two plans to consider:
Plan A: Run for 2 minutes the first day, and increase her running time by 30% every day.
Plan B: Run for 1 minute the first day and increase her running time by t minutes every day.

(i) Find the total time (in minutes) Penny has run after 20 days if she chooses Plan A.

(ii) Find the value of t such that the total time Penny runs in the first 20 days under both plans is the same.

9 The first term of an arithmetic progression is 18 and the common difference is d, where $d \neq 0$. The first term, the fourth term and the sixth term of this arithmetic progression are the first term, the second term and the third term, respectively, of a geometric progression with common ratio r.

(i) Write down two equations connecting d and r. Hence show that $r = \frac{2}{3}$ and find the value of d.

(ii) Find the sum to infinity of the geometric progression.

(iii) Find the value of n such that the sum of the first n terms of the arithmetic progression is zero.

4.3 Binomial expansions

1 Expand the following expressions using the binomial theorem.

(i) $(x+2)^4$

(ii) $(1-3x)^3$

2 Find the first three terms in the following expansions, fully simplifying each term.

(i) $(2x-3)^8$

(ii) $(3a-b)^7$

(iii) $\left(\frac{1}{x^2}+x^3\right)^7$

3 Find the coefficient of the x^2 term in the expansion of $\left(2x+\frac{1}{x}\right)^{10}$.

4 Find the term independent of x in the expansion of $\left(x^2 + \frac{5}{x^3}\right)^{15}$.

5 Find the coefficient of the x^2 term in the expansion of these expressions.

(i) $(2-3x)^6$

(ii) $(3+2x)(2-3x)^6$

6 Find the value of the constant b such that there is no term in x^3 in the expansion of $(1+bx)(x+2)^5$.

7 (i) Find the first three terms of $(3+u)^6$ in ascending powers of u.

(ii) Use the substitution $u = x - x^2$ in the answer to **(i)** to find the coefficient of the x^2 term in the expansion of $\left(3+x-x^2\right)^6$.

8 In the expansion of $(4 - ax)^7$ the coefficient of the x^3 term is -1120. Find the coefficient of the x^2 term.

9 **(i)** Find the first three terms, in ascending powers of x, in the expansion of $(3 - 2x^2)^7$.

(ii) Find the coefficient of the x^4 term in the expansion of $(1 + x^2)(3 - 2x^2)^7$.

10 The coefficient of the x^3 term in the expansion of $\left(k - \frac{1}{2}x\right)^6$ is 160. Find the value of the constant k.

Further practice

1 The third term of an arithmetic sequence is 12 and the sum of the first eight terms is 168.
Find the 14th term.

2 In an arithmetic progression, the first term is 25, the ninth term is 5 and the last term is -45.
Find the sum of all the terms in the progression.

3 The third term of a geometric progression is 128 and the seventh term is 40.5.
Find the fifth term and the sum to infinity.

4 An arithmetic sequence has a third term of 12 and a seventh term of 6.
Find the 21st term of the sequence.

5 The first three terms of a geometric sequence are $\sqrt{2},\ 2,\ \sqrt{8}$.
Find the common ratio, r, and the sixth term of the sequence.

6 The fifth and tenth terms of an arithmetic sequence are -40 and -20 respectively.
Find the value of n such that the sum of the first n terms is zero.

7 The second term in a geometric sequence is 9 and the fifth term is $1\frac{1}{8}$.
Find the sum to infinity of the sequence.

8 A car depreciates in value by 5% each year.
If the car was bought for \$45 000, find the value of the car eight years after it was bought.

9 Find the first three terms in the expansion of $\left(\frac{1}{x}+2x\right)^6$, fully simplifying each term.

10 Find the coefficient of the x^4 term in the expansion of $\left(\frac{x}{2}-\frac{3}{x}\right)^8$.

11 Find the term independent of x in the expansion of $\left(\frac{2}{x}+x^2\right)^{12}$.

12 Find the value of k such that the coefficient of the x^{-1} term in the expansion of $\left(kx+\frac{2}{x}\right)^5$ is 720.

13 The coefficient of the x^2 term in the expansion of $(3-x)^4+(2+ax)^5$, where a is a positive constant, is 554.
Find the value of a.

Past exam questions

1 **(i)** Find the coefficient of x in the expansion of $\left(2x-\frac{1}{x}\right)^5$. [2]

(ii) Hence find the coefficient of x in the expansion of $\left(1+3x^2\right)\left(2x-\frac{1}{x}\right)^5$. [4]

Cambridge International AS & A Level Mathematics 9709 Paper 12 Q1 June 2017

2 The sum of the 1st and 2nd terms of a geometric progression is 50 and the sum of the 2nd and 3rd terms is 30.
Find the sum to infinity. [6]

Cambridge International AS & A Level Mathematics 9709 Paper 11 Q5 November 2016

3 A water tank holds 2000 litres when full. A small hole in the base is gradually getting bigger so that each day a greater amount of water is lost.

(i) On the first day after filling, 10 litres of water are lost and this increases by 2 litres each day.

(a) How many litres will be lost on the 30th day after filling? [2]

(b) The tank becomes empty during the nth day after filling. Find the value of n. [3]

(ii) Assume instead that 10 litres of water are lost on the first day and that the amount of water lost increases by 10% on each succeeding day. Find what percentage of the original 2000 litres is left in the tank at the end of the 30th day after filling. [4]

Cambridge International AS & A Level Mathematics 9709 Paper 12 Q9 June 2016

4 (i) Write down the first 4 terms, in ascending powers of x, of the expansion of $(a - x)^5$. [2]

(ii) The coefficient of x^3 in the expansion of $(1 - ax)(a - x)^5$ is -200. Find the possible values of the constant a. [4]

Cambridge International AS & A Level Mathematics 9709 Paper 13 Q3 June 2015

5 (i) An arithmetic progression contains 25 terms and the first term is -15. The sum of all the terms in the progression is 525. Calculate

(a) the common difference of the progression, [2]

(b) the last term in the progression, [2]

(c) the sum of all the positive terms in the progression. [2]

(ii) A college agrees a sponsorship deal in which grants will be received each year for sports equipment. This grant will be \$4000 in 2012 and will increase by 5% each year. Calculate

(a) the value of the grant in 2022, [2]

(b) the total amount the college will receive in the years 2012 to 2022 inclusive. [2]

Cambridge International AS & A Level Mathematics 9709 Paper 12 Q10 November 2011

STRETCH AND CHALLENGE

1 Anna decides to reduce the time she spends on Facebook by the same number of minutes each week. In week 7 of her plan she uses Facebook for 400 minutes. After 30 weeks the total amount of time she spent using Facebook was 1800 minutes.

For how many minutes did Anna use Facebook in week 1 of her plan?

2 The constant terms in the expansions of $\left(kx^3 - \frac{7}{x^3}\right)^6$ and $\left(kx^4 + \frac{m}{x^4}\right)^8$ are the same.

If k and m are positive constants, express k in terms of m.

3 The sum of the first two terms of a geometric progression is -3. The sum of the sixth and seventh terms is 729. Find the common ratio and the first term of the progression.

4 In an arithmetic progression the 12th term is 3 times the value of the 6th term and the sum of the first 40 terms is 148.

Find the common difference and the first term.

5 A stamp collector buys two stamps. The first is bought for \$55 000 and its value depreciates by \$2400 every year. The second stamp depreciates by 4% every year. After 10 years the stamps have the same value. Find the price the second stamp was bought for.

6 (i) For what values of the positive constant a is the coefficient of x^3 in the expansion of $(x + a)^{10}$ the largest coefficient?

(ii) Which values of the positive constant a make the coefficient of x^{43} the largest coefficient in the expansion of $(x + a)^{100}$?

5 Functions and transformations

5.1 The language of functions

1 Find the rule that links the x and $\mathrm{f}(x)$ values.

(i)

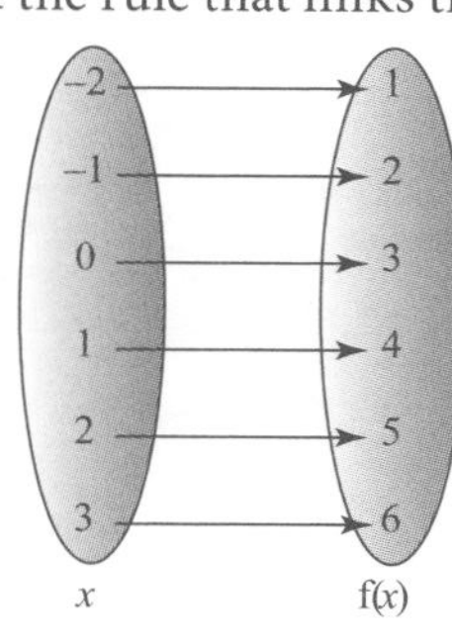

(ii)

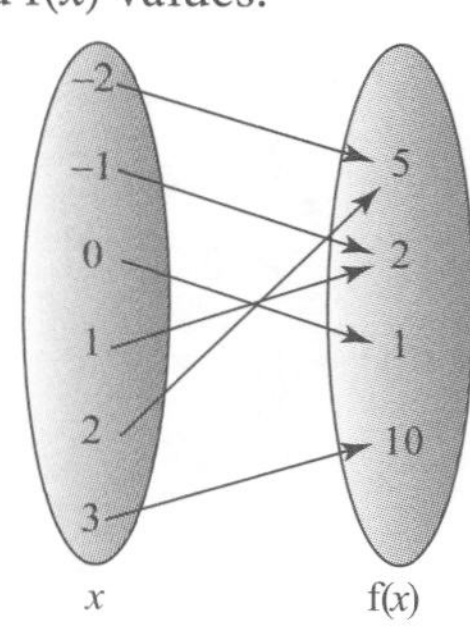

(iii)

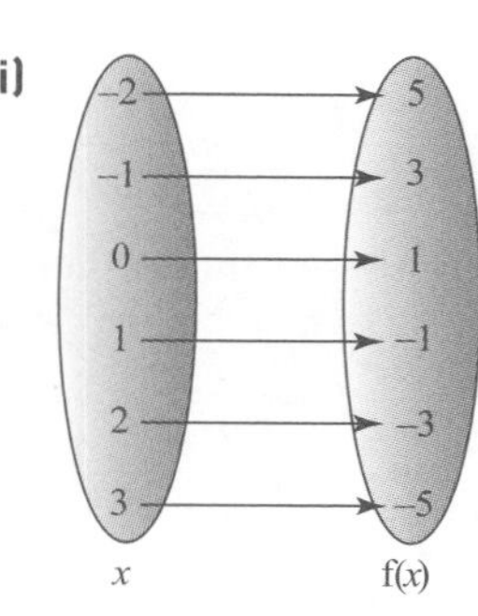

(iv)

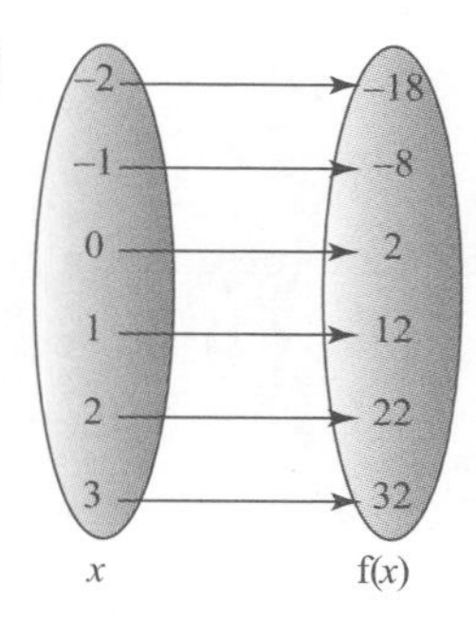

2 Which of the rules in question 1 are functions? Are any of the rules one–one functions?

3 Are the following functions? If so, are they one–one functions? Circle the correct answers.

(i)

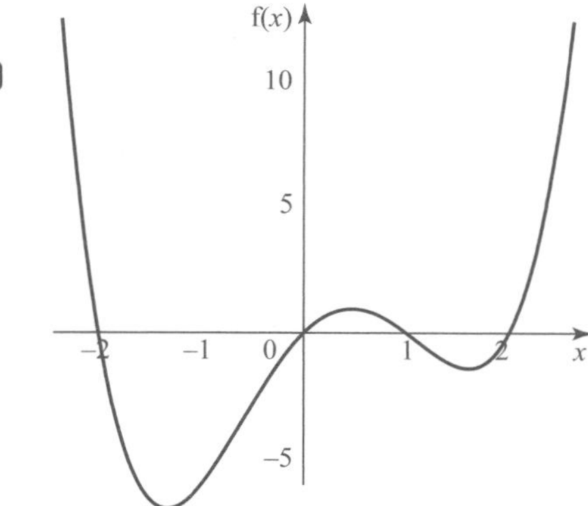

Function	Yes	No
One–one	Yes	No

(ii)

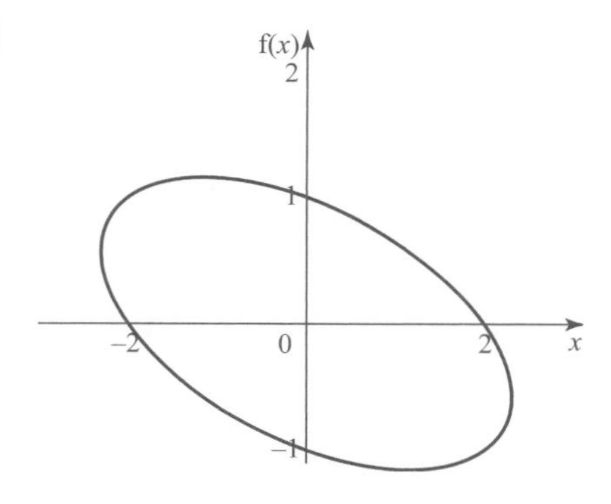

Function	Yes	No
One–one	Yes	No

(iii)

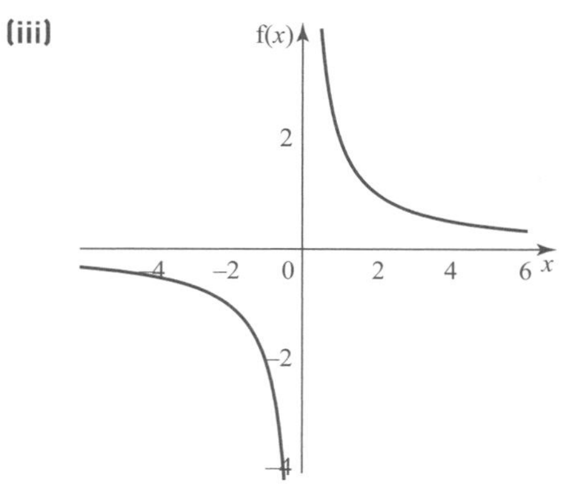

Function	Yes	No
One–one	Yes	No

(iv)

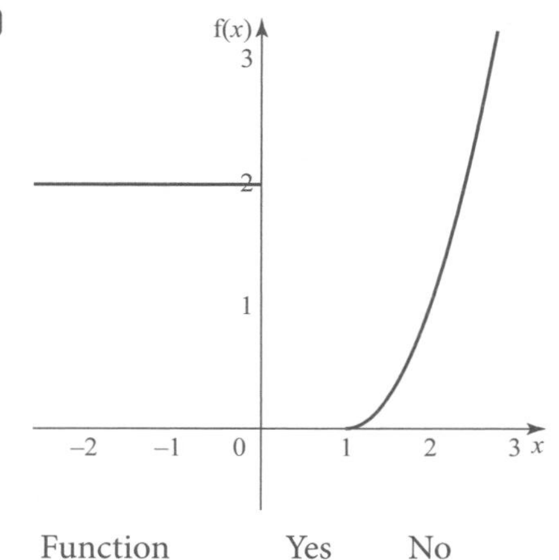

Function	Yes	No
One–one	Yes	No

4 For the function $f(x) = x^2 - 2x + 1$,

(i) find $f(-2)$

(ii) find and simplify $f(x + 1)$.

5 Find the domain and range of these functions.

(i) $f(x) = (x - 2)^2 - 2$

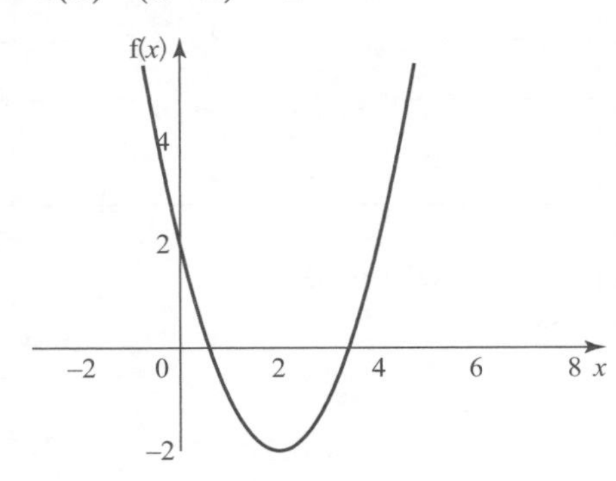

Domain ______________________

Range ______________________

(ii) $f(x) = -(x + 1)^2 + 3$

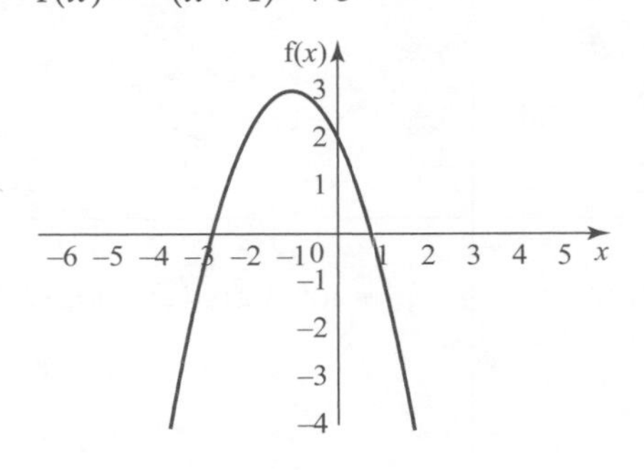

Domain ______________________

Range ______________________

(iii) $f(x) = \sqrt{x - 1} + 2$

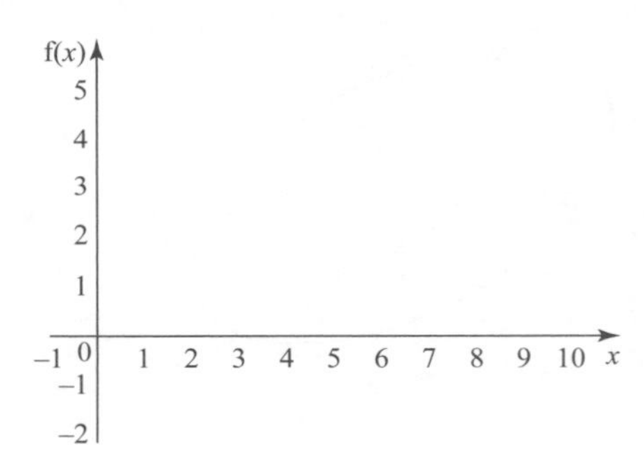

Domain ______________________

Range ______________________

(iv) $f(x) = (x + 1)^3$

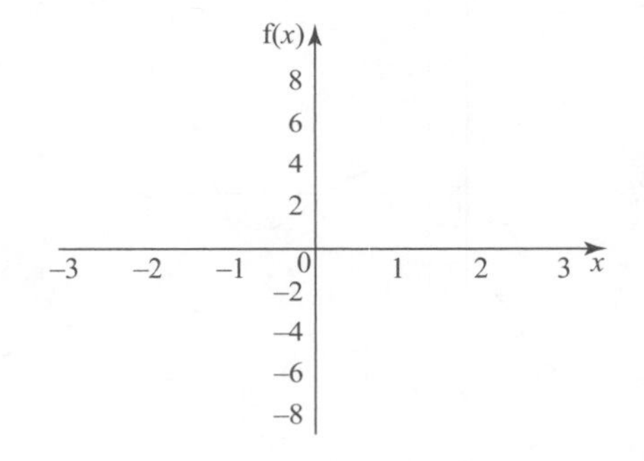

Domain ______________________

Range ______________________

6 A function is defined by $f(x) = x^2 - 4x + 1$, $x \leqslant 1$. Find the range.

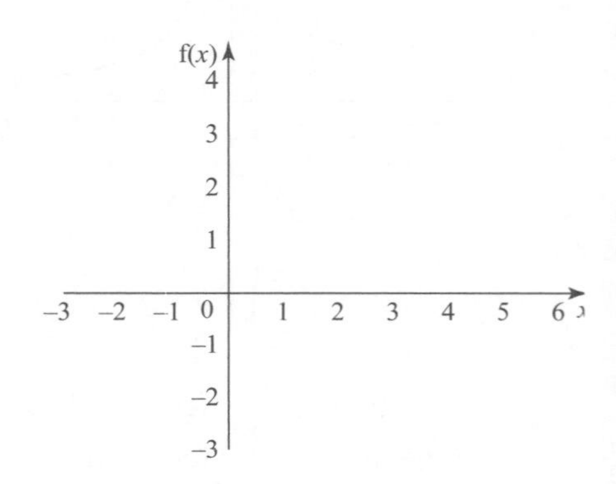

7 Find the domain and range of $g(x) = \sqrt{2 - \sqrt{x}}$.

8 Find the domain and range of $f(x) = 2x - x^2$.

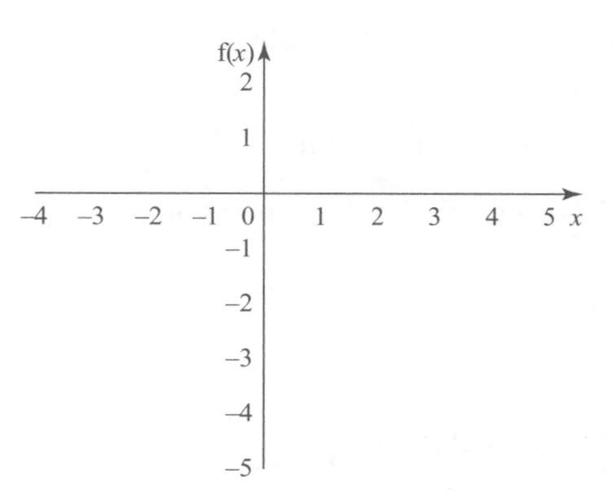

9 Find the range of $k(x) = 2x^2 + 8x + 3$ given the domain $x \geqslant -1$.

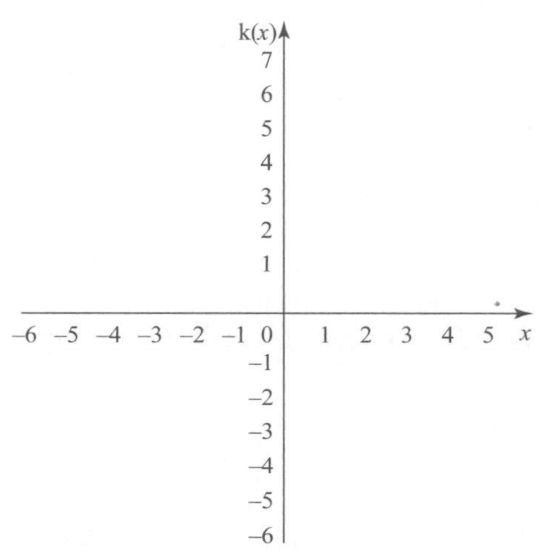

5.2 Composite functions

1 For the functions $f(x) = 1 - x$ and $g(x) = 1 - x^2$, find the following.

(i) $f(-2)$ **(ii)** $g(-2)$ **(iii)** $fg(-2)$ **(iv)** $gf(-2)$

(v) $fg(x)$ **(vi)** $gf(x)$ **(vii)** $ff(x)$ **(vii)** $gg(x)$

2 The function f is defined as $f: x \mapsto ax + b$ for $x \in \mathbb{R}$ where a and b are constants. It is given that $ff(x) = 9x - 4$. Find the possible values of a and b.

3 The functions f and g are defined by

$$f: x \mapsto 3x + 2$$
$$g: x \mapsto 4 + \frac{2}{x} \text{ for } x \neq 0$$

(i) Find and simplify the expression for $fg(x)$.

(ii) Show that $gf(x) = \frac{12x + 10}{3x + 2}$.

(iii) Solve $ff(x) = -3$.

(iv) Find the value(s) of k such that the equation $g(x) = kx$ has two solutions.

4 Find the values of a and b for the function $g: x \mapsto b - ax$ given that $g(1) = -1$ and $gg(1) = 5$.

5 Functions f and g are defined by

$$f: x \mapsto 3(x + 2) \quad \text{for } x \in \mathbb{R}$$
$$g: x \mapsto x^2 + 4x \quad \text{for } x \geqslant -2$$

(i) Find the set of values of x which satisfy $fg(x) \geqslant g(x)$.

(ii) Find the set of values of x which satisfy $gf(x) \leqslant 45$.

5.3 Inverse functions

1 Find an expression for the inverse of each of the following functions.

On the axes below, sketch the graphs of f(x) and $f^{-1}(x)$, showing the coordinates of their point of intersection and the relationship between the graphs.

(i) $f(x) = 2x + 1$ **(ii)** $f(x) = 5 - x$

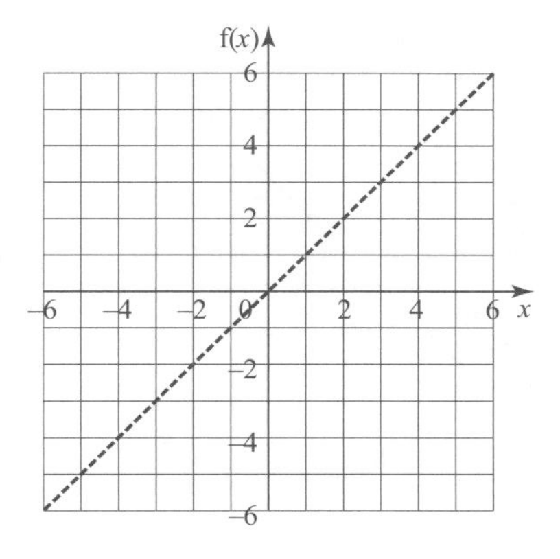

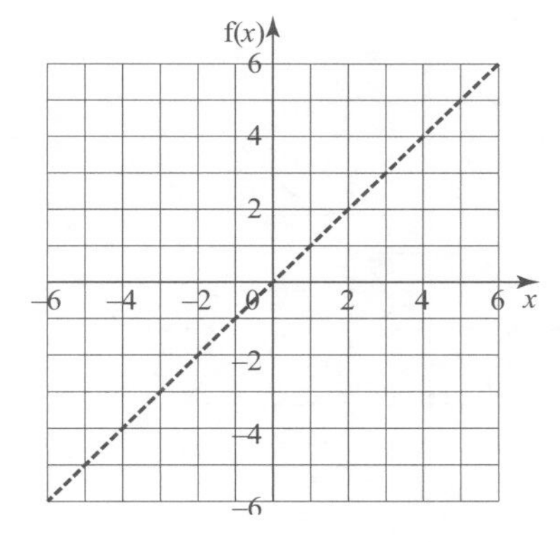

(iii) $f(x) = 1 - 5x$ **(iv)** 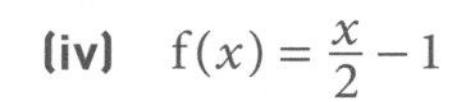 $f(x) = \frac{x}{2} - 1$

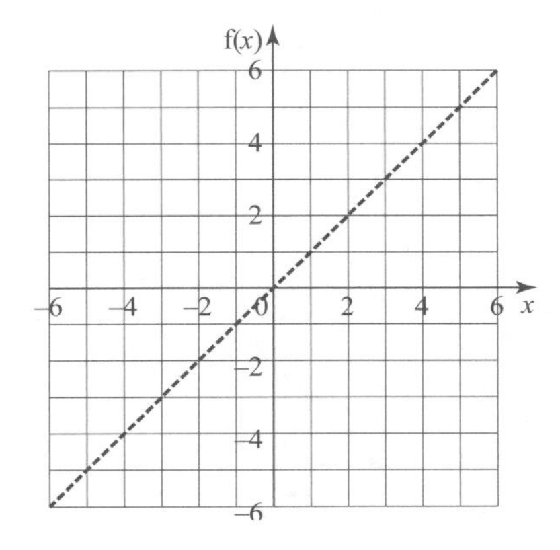

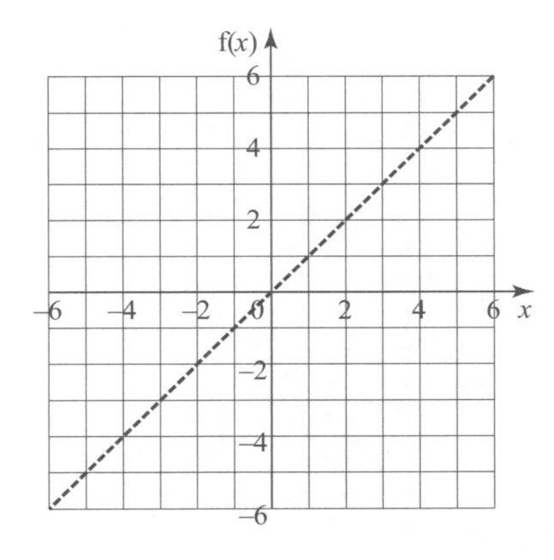

2 For each of the following quadratic functions:

- find the *least* value of k such that the domain $x \geqslant k$ means the function has an inverse
- sketch the graph of the function and its inverse on the axes for the domain $x \geqslant k$
- find an expression for the inverse of the function
- state the domain and range of the function and its inverse.

(i) $f(x)=x^2-4$

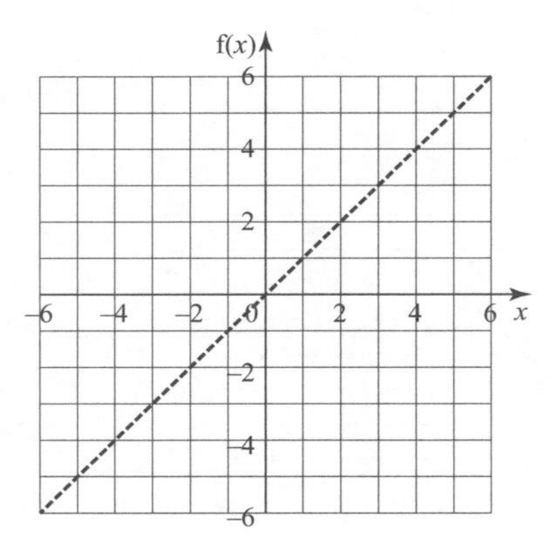

Domain $f(x)$: $x \geqslant$ ____________

Range $f(x)$: ____________

Domain $f^{-1}(x)$: ____________

Range $f^{-1}(x)$: ____________

(ii) $f(x)=x^2-4x$

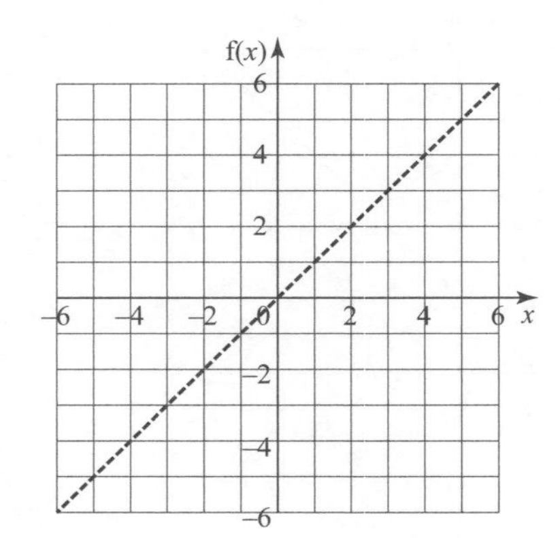

Domain $f(x)$: $x \geqslant$ ____________

Range $f(x)$: ____________

Domain $f^{-1}(x)$: ____________

Range $f^{-1}(x)$: ____________

(iii) $f(x)=x^2+2x-4$

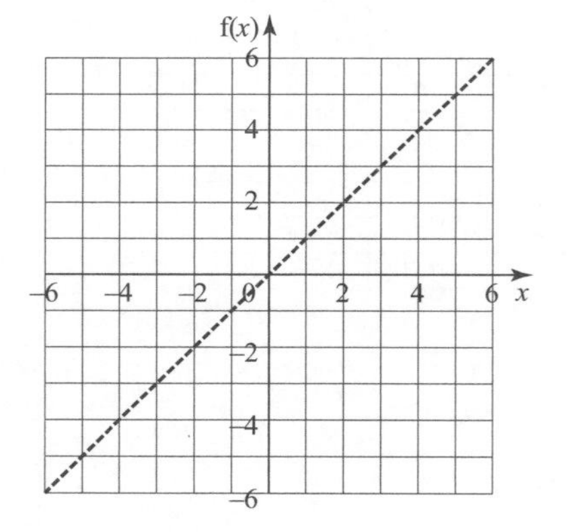

Domain $f(x)$: $x \geqslant$ ____________

Range $f(x)$: ____________

Domain $f^{-1}(x)$: ____________

Range $f^{-1}(x)$: ____________

(iv) $f(x)=2x^2+8x+7$

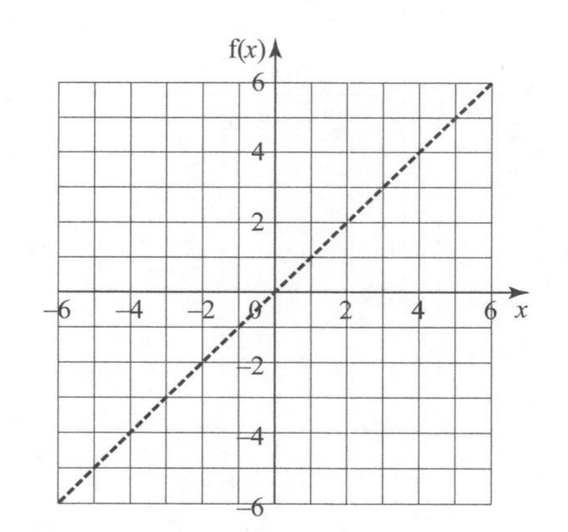

Domain $f(x)$: $x \geqslant$ ____________

Range $f(x)$: ____________

Domain $f^{-1}(x)$: ____________

Range $f^{-1}(x)$: ____________

3 The function $g : x \mapsto x^2 + 3x + 1$ has an inverse when $x \leqslant k$. Find the *largest* possible value of k. Sketch the graph of $g(x)$ for $x \leqslant k$ and the inverse.

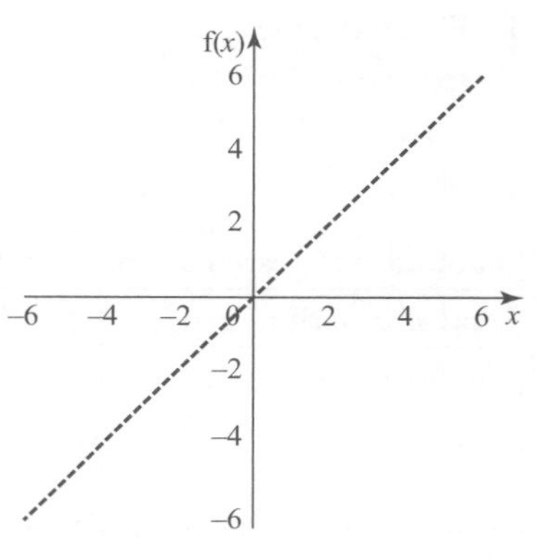

4 Find an expression for the inverse of the function $f(x) = 2(x+1)^3 - 1$.

The graphs of f(x) and its inverse are shown below.

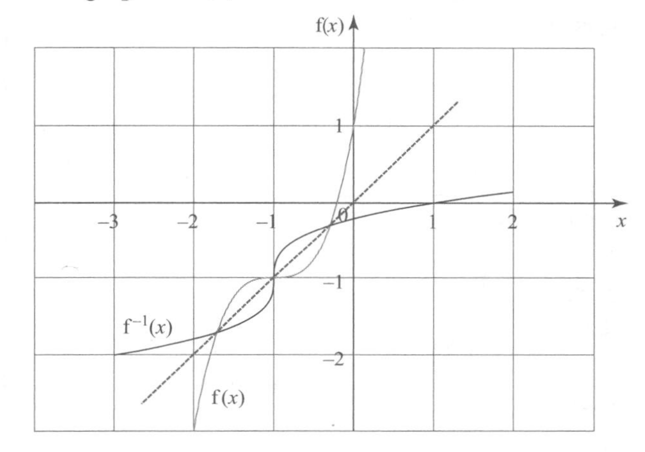

5 The graph of $f(x) = \cos x$ is shown for $0 \leqslant x \leqslant 2\pi$.
What is the maximum value of k such that $0 \leqslant x \leqslant k$ means that f(x) is one–one?

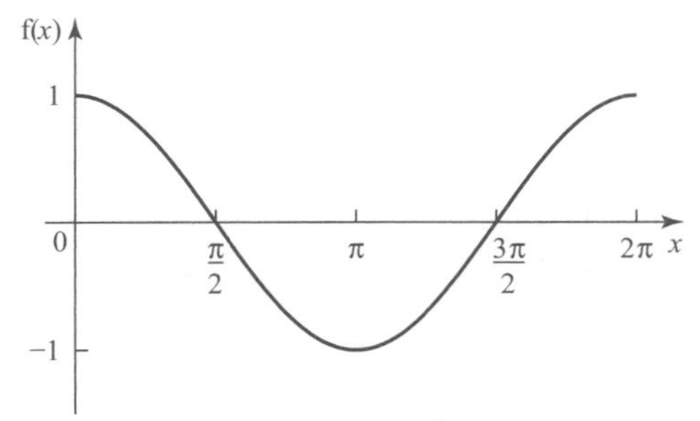

5.4 Transformations

1 Fill in the blank spaces in the sentences in the table below with the correct words from the following list: *translate, horizontally, vertically, upwards, downwards, right, left, further away, closer, horizontally stretch, vertically stretch, reflect, x-axis, y-axis, vector.*
Sketch an example graph to show your understanding.

Summary of graphical transformations on $y = \mathrm{f}(x)$		
Transformation	**Detail**	**Example graph**
For $y = \mathrm{f}(x) + b$, the effect of b is to ______ the graph of $y = \mathrm{f}(x)$ ______ through b units.	• If $b > 0$ it moves ______ • If $b < 0$ it moves ______	
For $y = \mathrm{f}(x - a)$, the effect of a is to ______ the graph of $y = \mathrm{f}(x)$ ______ through a units.	• If $a > 0$ it moves to the ______ • If $a < 0$ it moves to the ______	
For $y = \mathrm{f}(x - a) + b$, the graph of $y = \mathrm{f}(x)$ has been ______ ______ by a units and ______ ______ by b units.	We say it has been translated by the vector $\begin{pmatrix} a \\ b \end{pmatrix}$.	
For $y = p\mathrm{f}(x)$, $p > 0$, the effect of p is to ______ ______ the graph of $y = \mathrm{f}(x)$ by a factor of p.	• If $p > 1$ it moves points on $y = \mathrm{f}(x)$ ______ from the x-axis. • If $0 < p < 1$ it moves points on $y = \mathrm{f}(x)$ ______ to the x-axis.	
For $y = \mathrm{f}(kx)$, $k > 0$, the effect of k is to ______ ______ the graph of $y = \mathrm{f}(x)$ by a factor of $\frac{1}{k}$.	• If $k > 1$ it moves points on $y = \mathrm{f}(x)$ ______ to the y-axis. • If $0 < k < 1$ it moves points on $y = \mathrm{f}(x)$ ______ from the y-axis.	
For $y = -\mathrm{f}(x)$, the effect on the graph of $y = \mathrm{f}(x)$ is to ______ it in the ______.		
For $y = \mathrm{f}(-x)$, the effect on the graph of $y = \mathrm{f}(x)$ is to ______ it in the ______.		

2 The diagram shows the graph of $y = \mathrm{f}(x)$. Sketch the graph of $y = \mathrm{f}(x + 1) - 2$ on the grid.

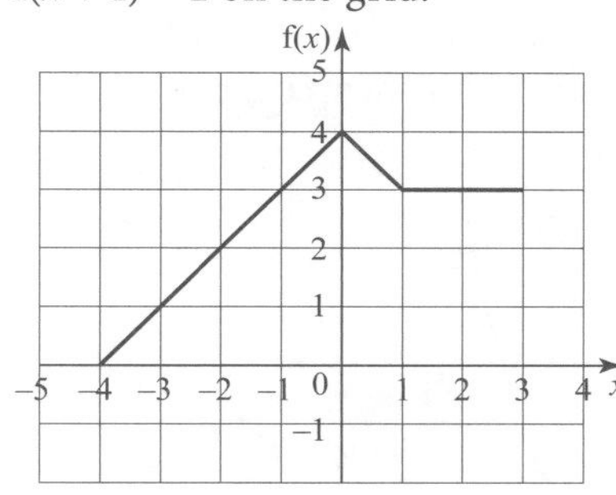

3 **(i)** The graph shows $y = \mathrm{f}(x)$ where $\mathrm{f}(x) = (x - 1)^2$. Sketch the graphs of the following on the grid.

(a) $y = -\mathrm{f}(x)$ **(b)** $y = \mathrm{f}(-x)$ **(c)** $y = -\mathrm{f}(-x)$.

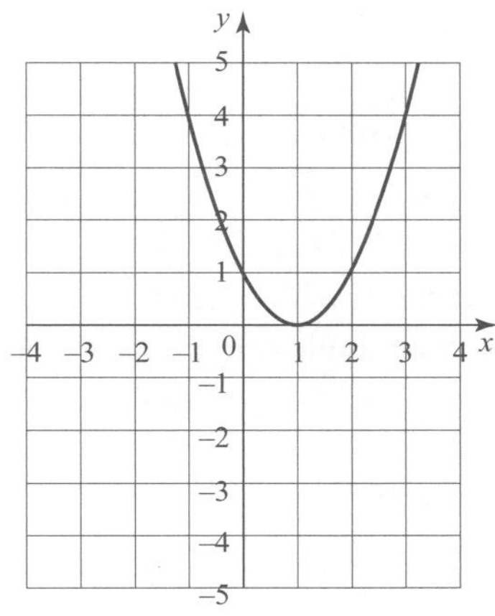

(ii) The curve $y = (x - 1)^2$ is translated by $\begin{pmatrix} 0 \\ 2 \end{pmatrix}$. Find and simplify the equation of the translated curve.

4 For the graph of $y = \mathrm{f}(x)$ below sketch on the grid the graphs of the following.

(i) $y = \mathrm{f}(x) + 2$ **(ii)** $y = -\mathrm{f}(x)$ **(iii)** $y = \mathrm{f}(x - 1)$

(iv) $y = \mathrm{f}(2x)$ **(v)** $y = 2\mathrm{f}(x)$ **(vi)** $y = \mathrm{f}(-x)$

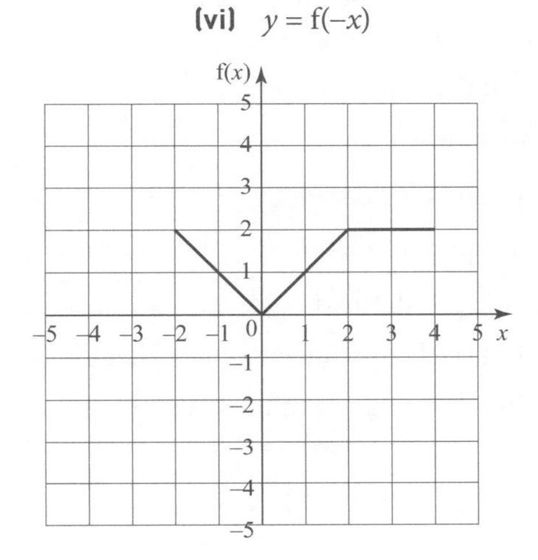

5 The graph of $f(x) = x^2 - 3x$ is translated 3 units to the right and 1 unit up (i.e. by the vector $\begin{pmatrix} 3 \\ 1 \end{pmatrix}$).
Find the equation of the new curve.

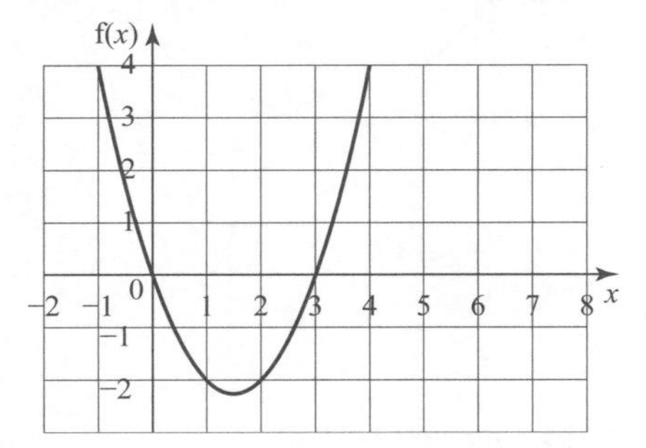

6 The graph of $y = f(x)$ is transformed to the graph of $y = 2f(-x)$.
Describe fully the two single transformations which have been combined to give the resulting transformation.

Further practice

1 A function f is defined by $f : x \mapsto x^2 - 4x + 2$ for $x \geqslant 2$. Find the domain and range of $f^{-1}(x)$.

2 The function f is defined by $f : x \mapsto \dfrac{x+2}{4x-1}$, $x \in \mathbb{R}$, $x \neq \dfrac{1}{4}$.

(i) Show that $ff(x) = x$.

(ii) Hence, or otherwise, obtain an expression for $f^{-1}(x)$.

3 A function f is defined by $f : x \mapsto ax + b$ for $x \in \mathbb{R}$, where a and b are constants. It is given that $f(-1) = 1$ and $f(2) = 7$.

(i) Find the values of a and b.

(ii) Solve $ff(x) = 1$.

4 The function j is defined by $j : x \mapsto \dfrac{2}{3-x}$ for $x \in \mathbb{R}$, $x \neq 3$.
Find an expression for $j^{-1}(x)$, the inverse of j.

5 A function g is defined by $g : x \mapsto x^2 + 6x + 2$, $x \geqslant k$.

(i) Find the least value of k for which $g(x)$ has an inverse.

(ii) Find the equation of the inverse in this case.

6 Functions f and g are defined by

$f: x \mapsto 2-3x,\ x \in \mathbb{R},\ x \geqslant 0$

$g: x \mapsto \dfrac{x+2}{2x+7},\ x \in \mathbb{R},\ x \neq -\dfrac{7}{2}$

(i) Solve the equation $gf(x) = x$.

(ii) Express $f^{-1}(x)$ and $g^{-1}(x)$ in terms of x.

(iii) Show that the equation $g^{-1}(x) = x - 4$ has no solutions.

(iv) Find and simplify an expression for $f^{-1}g(x)$.

(v) Sketch in a single diagram the graphs of $y = f(x)$ and $y = f^{-1}(x)$, making clear the relationship between the graphs.

7 The equation of a curve is $f(x) = x^3 + 1$. Find the equation of the curve when $f(x)$ is translated by the vector $\begin{pmatrix} -1 \\ 2 \end{pmatrix}$.

8 The graph of $y = f(x)$ is transformed to the graph of $y = -f(2x)$. Describe fully the two single transformations which have been combined to give the resulting transformation.

Past exam questions

1 The functions f and g are defined by

$f(x) = \dfrac{4}{x} - 2 \quad \text{for } x > 0,$

$g(x) = \dfrac{4}{5x+2} \quad \text{for } x \geqslant 0.$

(i) Find and simplify an expression for $fg(x)$ and state the range of fg. [3]

(ii) Find an expression for $g^{-1}(x)$ and find the domain of g^{-1}. [5]

Cambridge International AS & A Level Mathematics 9709 Paper 11 Q8 November 2016

2 The diagram shows the graph of $y = f^{-1}(x)$, where f^{-1} is defined by $f^{-1}(x) = \dfrac{1-5x}{2x}$ for $0 < x \leqslant 2$.

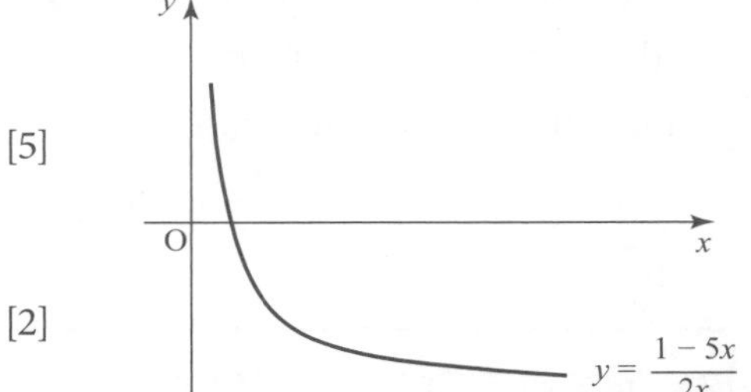

(i) Find an expression for $f(x)$ and state the domain of f. [5]

(ii) The function g is defined by $g(x) = \dfrac{1}{x}$ for $x \geqslant 1$. Find an expression for $f^{-1}g(x)$, giving your answer in the form $ax + b$, where a and b are constants to be found. [2]

Cambridge International AS & A Level Mathematics 9709 Paper 13 Q6 June 2015

3 The function f is defined by $f(x) = 3x + 1$ for $x \leqslant a$, where a is a constant. The function g is defined by

$g(x) = -1 - x^2 \quad \text{for } x \leqslant -1.$

(i) Find the largest value of a for which the composite function gf can be formed. [2]

For the case where $a = -1$,

(ii) solve the equation $fg(x) + 14 = 0$, [3]

(iii) find the set of values of x which satisfy the inequality $gf(x) \leqslant -50$. [4]

Cambridge International AS & A Level Mathematics 9709 Paper 13 Q8 November 2015

4 The functions f and g are defined for $x \geqslant 0$ by

$$f : x \mapsto 2x^2 + 3,$$
$$g : x \mapsto 3x + 2.$$

(i) Show that $gf(x) = 6x^2 + 11$ and obtain an unsimplified expression for $fg(x)$. [2]

(ii) Find an expression for $(fg)^{-1}(x)$ and determine the domain of $(fg)^{-1}$. [5]

(iii) Solve the equation $gf(2x) = fg(x)$. [3]

Cambridge International AS & A Level Mathematics 9709 Paper 12 Q8 March 2017

5 The diagram shows the function f defined for $0 \leqslant x \leqslant 6$ by

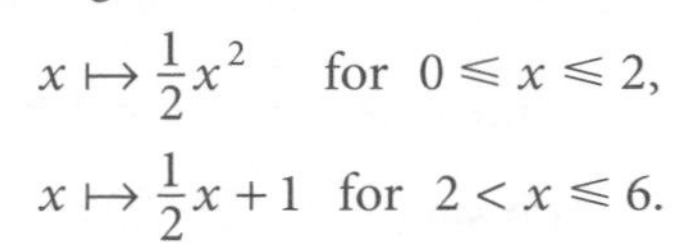

$$x \mapsto \frac{1}{2}x^2 \quad \text{for } 0 \leqslant x \leqslant 2,$$
$$x \mapsto \frac{1}{2}x + 1 \quad \text{for } 2 < x \leqslant 6.$$

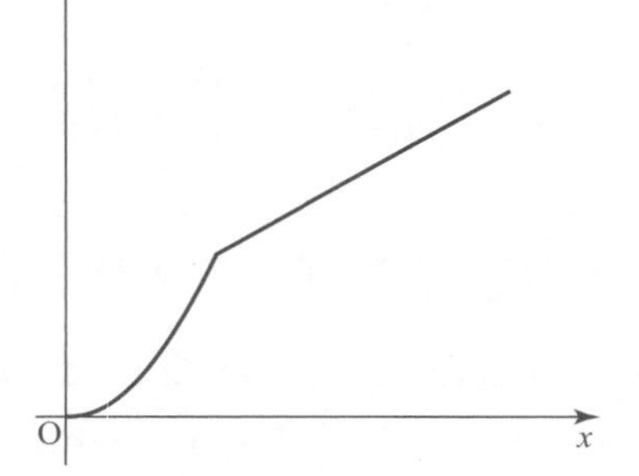

(i) State the range of f. [1]

(ii) On the diagram sketch the graph of $y = f^{-1}(x)$. [2]

(iii) Obtain expressions to define $f^{-1}(x)$, giving the set of values of x for which each expression is valid. [4]

Cambridge International AS & A Level Mathematics 9709 Paper 13 Q7 November 2010

6 The functions f and g are defined for $x \in \mathbb{R}$ by

$$f : x \mapsto 3x + a,$$
$$g : x \mapsto b - 2x,$$

where a and b are constants. Given that $ff(2) = 10$ and $g^{-1}(2) = 3$, find

(i) the values of a and b, [4]

(ii) an expression for $fg(x)$. [2]

Cambridge International AS & A Level Mathematics 9709 Paper 12 Q2 November 2011

STRETCH AND CHALLENGE

1 Functions f, g and h are defined as follows:

$$f(x) = 2x + 1$$
$$g(1) = f(0)$$
$$g(x) = fg(x - 1) \quad x \geqslant 1$$
$$h(0) = g(2)$$
$$h(x) = gh(x - 1) \quad x \geqslant 1$$

Find h(2).

2 A function f is defined by $f(x) = 3^x$. If $f(x + 1) + f(x - 1) = af(x)$, where a is a positive constant, find the value of a.

3 For a function $f(n) = (an + b)$, where a and b are integers, $f(2n - 1)$, $f(2n) - 1$ and $2f(n) - 1$ are three consecutive integers in some order.
Find all the possible functions $f(n)$.

6 Differentiation

6.1 Basic differentiation

1 Find the derivative of the following functions.

(i) $y = 10x^2 - 3x + 1$

(ii) $y = 4 + x - 5x^4$

(iii) $y = \frac{5}{x^2}$

(iv) $y = \frac{2x^3}{3}$

(v) $y = \frac{4}{x} - \frac{x}{4}$

(vi) $y = \frac{1}{4x^3}$

(vii) $y = 2\sqrt{x} - 3x$

(viii) $y = \sqrt[3]{x} + \frac{2}{3x}$

2 Find $f'(x)$ for each of these functions.

(i) $f(x) = \frac{2x+1}{x^2}$

(ii) $f(x) = 6\sqrt[3]{x^5}$

3 Find $f'(9)$ if $f(x) = \frac{6}{\sqrt{x}} + 5x$.

4 Find the coordinates of the point on the curve $y = 3 - 8x - x^2$ where the gradient is 2.

5 The curve $f(x) = x^3 + x^2 + 2x - 1$ has two points where the gradient of the curve is 3. Find the coordinates of the points.

6 The points A(0, 2), B(1, 7), C(1.4, 9.56), D(1.9, 13.21) and E(2, 14) lie on the curve $y = f(x)$. The table below shows the gradients of the chords AE and BE.

(i) Complete the table to show the gradients of CE and DE.

Chord	AE	BE	CE	DE
Gradient of chord	6	7		

(ii) State what the values in the table indicate about the value of $f'(2)$.

6.2 Tangents and normals

1 Find the equation of the tangent to the curve $y = 2x^4 + x^2 - 1$ at $(-1, 2)$.

2 Find the equation of the normal to the curve $g(x) = \frac{4}{\sqrt{x}} - 3$ at $x = 4$.

3 **(i)** Show that the equation of the tangent to the curve $h(x) = 6x - x^2$ at the point P(2, 8) is $2x - y + 4 = 0$.

(ii) The normal to the curve at the point P meets the curve again at the point Q.
Find the coordinates of the point Q.

4 The gradient of the normal to the curve $y = x + \frac{k}{x}$ at the point where $x = -1$ is $\frac{1}{4}$.
Find the value of the constant k.

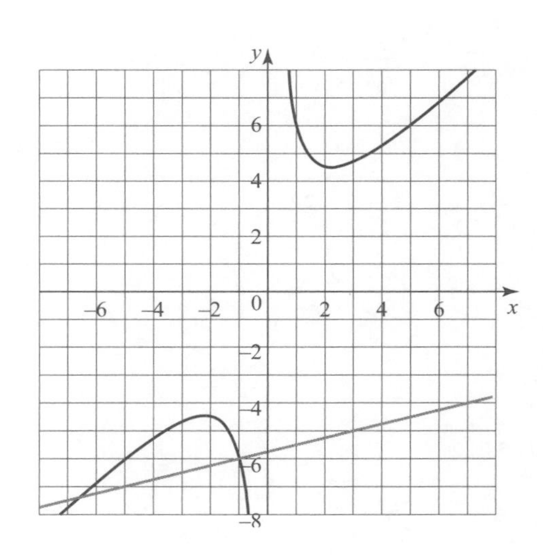

6.3 Maximum and minimum points and increasing and decreasing functions

1 Find the coordinates of the stationary points for the following curves, and determine their nature.

(i) $y = 2x^3 - 3x^2$ **(ii)** $f(x) = 4x + \frac{1}{x}$ **(iii)** $y = \sqrt{x} + \frac{1}{\sqrt{x}}$

2 Find the values of x where $g(x) = 3x^2 - 2x^3$ is increasing.

3 Show that the function $y = -x + x^2 - x^3$ is decreasing for all $x \in \mathbb{R}$.

4 Sketch the graphs of the following functions. Use any information found in previous questions to help you.

(i) $y = 2x^3 - 3x^2$

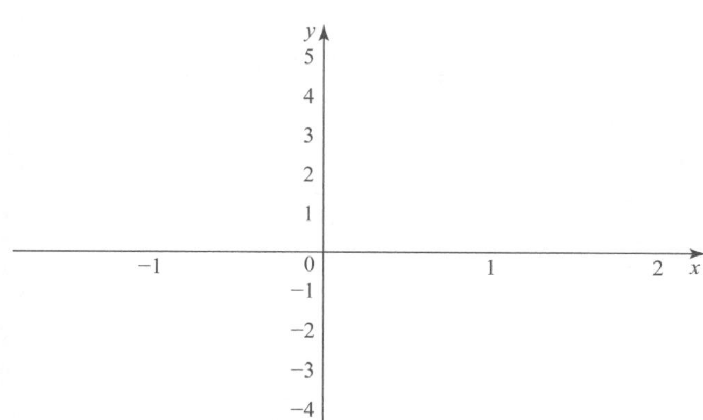

(ii) $g(x) = 3x^2 - 2x^3$

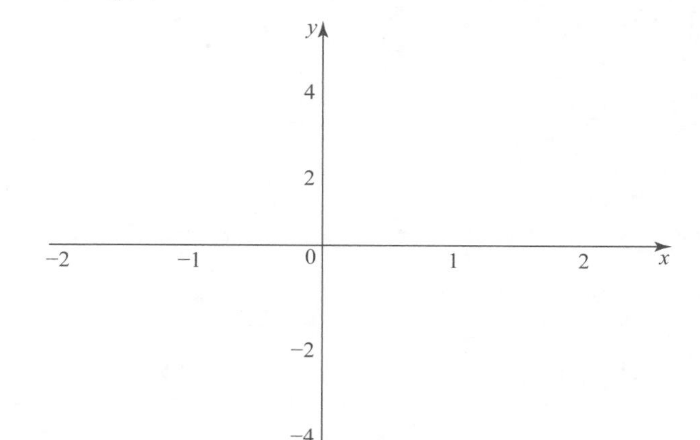

5 A curve $y = f(x)$ has a stationary point at the point $(1, -9)$. It is given that $f'(x) = 3x^2 + kx - 8$, where k is a constant.

(i) Find the value of k.

(ii) Hence find the x-coordinate of the other stationary point on the curve.

(iii) Find $f''(x)$ and determine the nature of each of the stationary points on $y = f(x)$.

6.4 The chain rule

1 Use the chain rule to find $\frac{dy}{dx}$ in each case.

(i) $y = (x - 4)^7$

(ii) $y = (5 - 2x)^9$

(iii) $y = \sqrt{3 + 4x}$

(iv) $y = \frac{2}{x + 3}$

(v) $y = \frac{4}{(1 - 2x)^2}$

2 Find the equation of the tangent to $f(x) = 3\left(1 - \frac{x}{6}\right)^4$ where $x = 12$.

3 Find the coordinates of the stationary point(s) of the function $y = \frac{8}{x^2 - 4x}$.

4 The function g is defined by $g : x \rightarrow 3(x + 2)^3 - 5, x > -2$.
Obtain an expression for $g'(x)$ and use your answer to explain why g has an inverse.

5 A curve has the equation $y = \frac{2}{x-1} + 2x$.

(i) Find $\frac{dy}{dx}$ and $\frac{d^2y}{dx^2}$.

(ii) Find the coordinates of the maximum point A and the minimum point B on the curve.

6.5 Applications

1 Two numbers have a sum of 18. Find the maximum value of the product of the numbers.

2 A farmer wants to construct three pens for his chickens as shown.
Each pen has the same width, x.
The farmer has 36 m of material to make the pens.
Find the dimensions of each pen so that the area of each pen is a maximum.

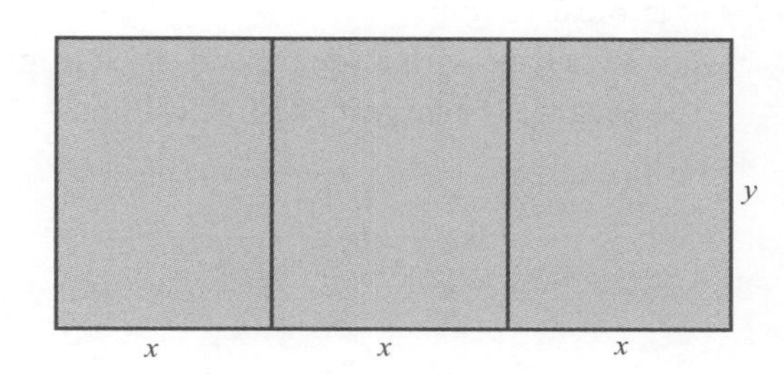

3 A 5 litre paint can has radius r and height h. (5 litres = 5000 cm^3)

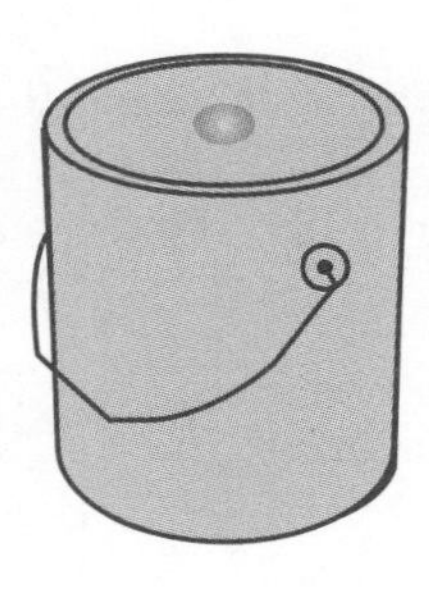

(i) Show that the surface area of the can, S, is given by

$$S = \frac{10\,000}{r} + 2\pi r^2$$

(ii) Given that r can vary, find the value of r for which S has a stationary value.

(iii) Determine whether this stationary value is a maximum or a minimum.

4 The post office will accept a package for shipment only if the sum of the girth (the distance around the middle) and the length is at most 120 cm.
For a package with a square end, what dimensions will give the package the largest possible volume?

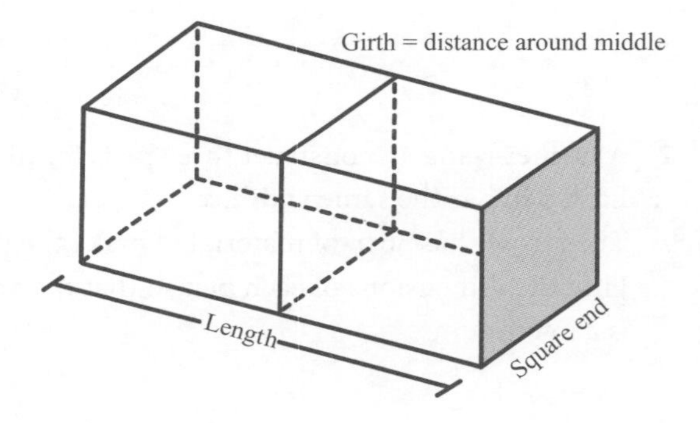

5 An open box is being constructed whose base length is 4 times the base width and whose volume is 50 m^3.
If the materials used to build the box cost \$12 per square metre for the bottom and \$5 per square metre for the sides, what are the dimensions of the least expensive box?

6.6 Rates of change

1 The area of a square is increasing at a rate of $8\,\text{cm}^2$ per second.
Find the rate at which the length of the side is increasing at the instant when the side length is 20 cm.

2 Water is being pumped into a cylindrical tank of radius 1.2 m at a rate of $0.5\,\text{m}^3$ per minute.
Find the rate at which the water level is rising.

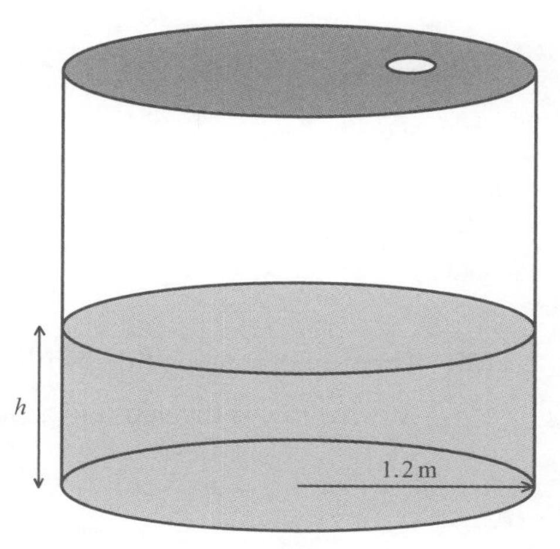

3 The equation of a curve is $y=(0.2x-1)^{10}$. A point with coordinates (x, y) moves along the curve in such a way that the rate of increase of x has the constant value 0.08 units per second.
Find the rate of increase of y at the instant when $x=10$.

4 A horse is training on a circular track of radius 200 m as shown. The horse starts at the point A and gallops to B at a constant 10 metres per second. It is followed by a camera at the centre of the track at O.
Find the rate of change of the angle θ during this time, in radians per second.

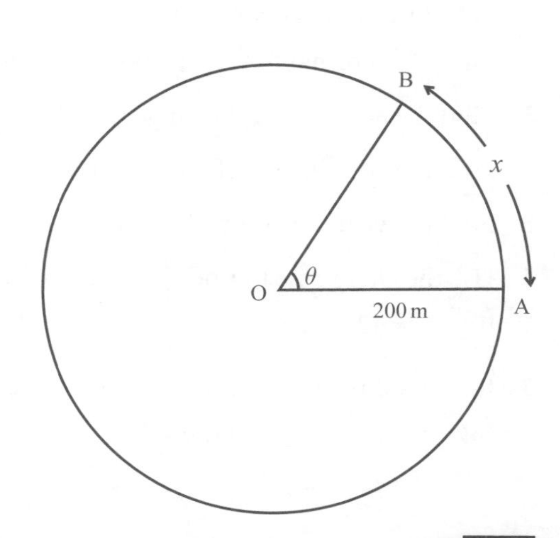

5 The diagram shows a water trough.

(i) When the depth of water in the trough is h cm, show that the trough contains $\frac{45}{2}h^2 + 3600h$ cm^3.

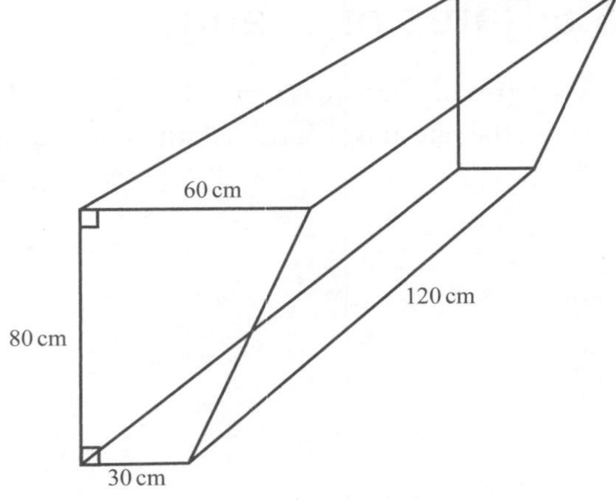

(ii) What is the depth of the water when the trough contains 40 litres?

(iii) The trough is being filled with a hose at a rate of 4 litres per minute.
At what rate is the depth increasing when there is 40 litres of water in the trough?

Further practice

1 Find the coordinates of the point on the curve $y = \frac{1}{3}x^3 + \frac{1}{2}x^2 - 3x$ where the gradient is -1.

2 **(i)** The curve C has equation $y = \frac{1}{x^2} + x$. The tangent to C at the point P(1, 2) intersects the x-axis at the point Q and the y-axis at the point R. Find the length of PR.

(ii) The normal to C at the point P intersects the curve C again at the point S. Find the coordinates of S.

3 Find the equation of the tangent and normal to the curve $f(x) = x + \frac{4}{x}$ at $x = 1$.

4 Find the coordinates of the stationary points for the curve $y = x^4 - 8x^2 + 2$, and determine their nature.

5 Find the values of x for which the function $f(x) = 6x^2 - x^3$ is increasing.

6 Use the chain rule to find $\frac{dy}{dx}$ in each case.

(i) $y = (3x + 2)^8$ **(ii)** $y = \sqrt[3]{1 - 9x}$ **(iii)** $y = \frac{5}{\sqrt{5x + 3}}$

7 **(i)** Find the value of the constant k such that the curve $y = \frac{k + 1}{2x + 3} + x$ has a gradient of 2 when $x = -1$.

(ii) When $k = 1$, find the values of x where the function is increasing.

8 An athletics track has a rectangular centre with semicircular area at each end.
The perimeter of the track is 400 m.

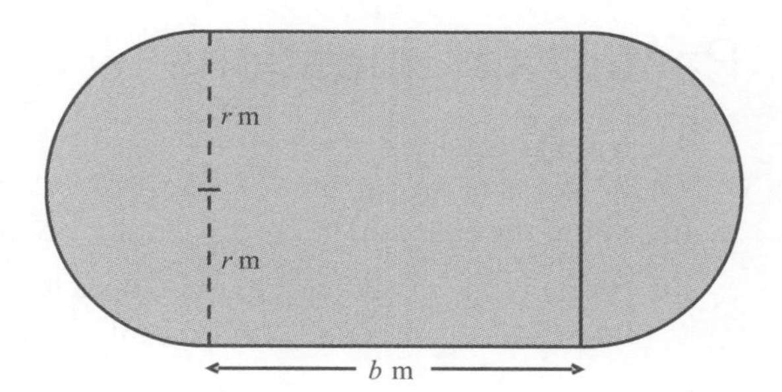

(i) Show that $b = 200 - \pi r$.

(ii) Find an expression for the area, A, of the rectangular section of the track, and hence find the values of r and b that maximise this area.

9 The diagram shows a rectangle inscribed in a right-angled isosceles triangle with a hypotenuse of length 2.
What is the largest area that the rectangle can have?

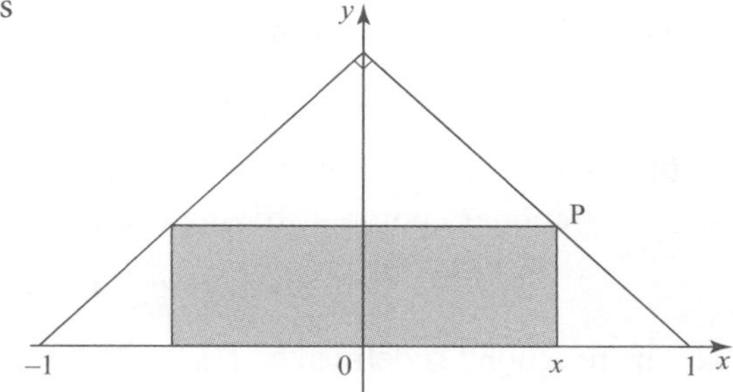

10 The ice cube shown is melting so that the rate of decrease of the sides of the cube is 5 mm per minute.
Find the rate of decrease of the surface area of the ice cube when the side length l is 2 cm.

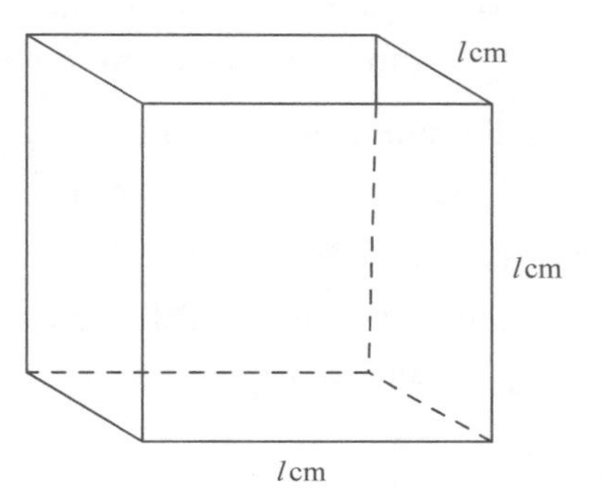

11 The equation of a curve is $y = \sqrt{1 + 6x}$. A point with coordinates (x, y) moves along the curve in such a way that the rate of increase of x has the constant value 0.02 units per second.
Find the rate of increase of y at the instant when $x = 4$.

12 The diagram shows a saltshaker which consists of a cylinder with radius r cm and height h cm and a hemisphere with radius r on top.
The shaker has a volume of 100 cm^3.

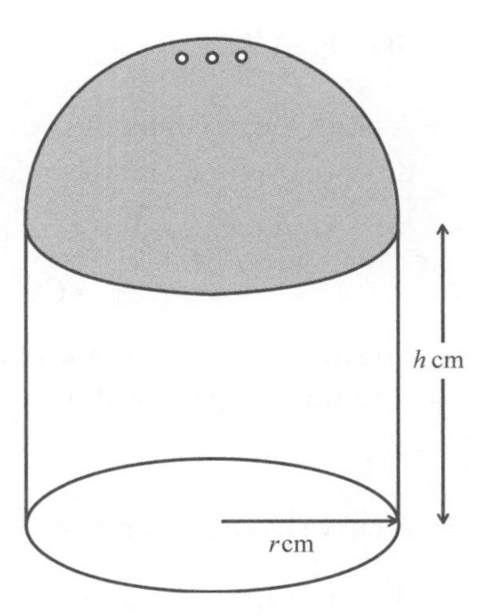

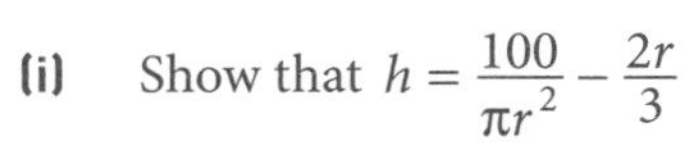

(i) Show that $h = \dfrac{100}{\pi r^2} - \dfrac{2r}{3}$

(ii) Show that the surface area, S, of the shaker is given by $S = \dfrac{200}{r} + \dfrac{5\pi}{3}r^2$.

(iii) Given that r can vary, find the value of r for which S has a stationary value.

(iv) Determine whether this stationary value is a maximum or a minimum.

Past exam questions

1 A curve has equation $y = 3 + \frac{12}{2-x}$.

(i) Find the equation of the tangent to the curve at the point where the curve crosses the x-axis. [5]

(ii) A point moves along the curve in such a way that the x-coordinate is increasing at a constant rate of 0.04 units per second. Find the rate of change of the y-coordinate when $x = 4$. [2]

Cambridge International AS & A Level Mathematics 9709 Paper 12 Q5 June 2017

2 The point P(3, 5) lies on the curve $y = \frac{1}{x-1} - \frac{9}{x-5}$.

(i) Find the x-coordinate of the point where the normal to the curve at P intersects the x-axis. [5]

(ii) Find the x-coordinate of each of the stationary points on the curve and determine the nature of each stationary point, justifying your answers. [6]

Cambridge International AS & A Level Mathematics 9709 Paper 11 Q11 November 2016

3 The function f is defined by $f(x) = 2x + (x+1)^{-2}$ for $x > -1$.

(i) Find $f'(x)$ and $f''(x)$ and hence verify that the function f has a minimum value at $x = 0$. [4]

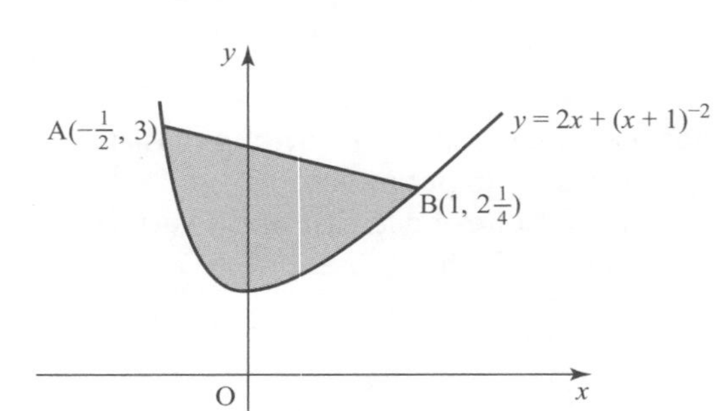

The points $A(-\frac{1}{2}, 3)$ and $B(1, 2\frac{1}{4})$ lie on the curve $y = 2x + (x+1)^{-2}$, as shown in the diagram.

(ii) Find the distance AB. [2]

(iii) Find, showing all necessary working, the area of the shaded region. [6]

Cambridge International AS & A Level Mathematics 9709 Paper 13 Q10 November 2015

4 The base of a cuboid has sides of length x cm and $3x$ cm. The volume of the cuboid is 288 cm^3.

(i) Show that the total surface area of the cuboid, A cm^2, is given by

$$A = 6x^2 + \frac{768}{x}.$$ [3]

(ii) Given that x can vary, find the stationary value of A and determine its nature. [5]

Cambridge International AS & A Level Mathematics 9709 Paper 13 Q9 May 2014

5 A curve has equation $y = \frac{12}{3-2x}$.

(i) Find $\frac{dy}{dx}$. [2]

A point moves along this curve. As the point passes through A, the x-coordinate is increasing at a rate of 0.15 units per second and the y-coordinate is increasing at a rate of 0.4 units per second.

(ii) Find the possible x-coordinates of A. [4]

Cambridge International AS & A Level Mathematics 9709 Paper 12 Q4 November 2014

6 The non-zero variables x, y and u are such that $u = x^2y$. Given that $y + 3x = 9$, find the stationary value of u and determine whether this is a maximum or a minimum value. [7]

Cambridge International AS & A Level Mathematics 9709 Paper 13 Q6 June 2013

7 A curve has equation $y = \frac{k^2}{x+2} + x$, where k is a positive constant. Find, in terms of k, the values of x for which the curve has stationary points and determine the nature of each stationary point. [8]

Cambridge International AS & A Level Mathematics 9709 Paper 13 Q9 November 2013

STRETCH AND CHALLENGE

1 The normal to the curve $g(x) = kx - x^2$ at $x = 2$ has a gradient of $\frac{1}{2}$.

(i) Find the value of k.

(ii) The normal intersects the curve again at the point P. Find the coordinates of P.

2 The positions of a submarine (S) and shipwreck (W) are shown in the diagram. Due to territorial issues, the closest the submarine can get to the shipwreck is 30 km horizontally from its position at the bottom of the ocean. The shipwreck is 15 km below the surface. For the first 10 km in depth, the submarine can travel at 10 km/h but for the last 5 km due to the building pressure it can only travel at 5 km/h.

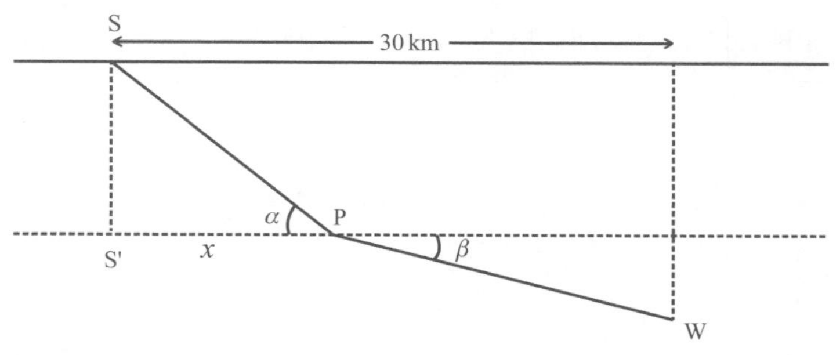

Rather than go directly to W from S, the submarine goes to a point P on the line where the speed must change, then changes course. The distance S′P is x.

(i) Find expressions for $\cos\alpha$ and $\cos\beta$ in terms of x.

(ii) Show that the minimum time to get from S to W happens when $\cos\alpha = 2\cos\beta$ (you do not need to show it is a minimum).

3 (i) A cylinder is enclosed in a sphere of radius R. Find the dimensions r and h, in terms of R, of the cylinder with maximum volume.

(ii) Find the ratio of the volume of the sphere to the cylinder with maximum volume in its simplest form.

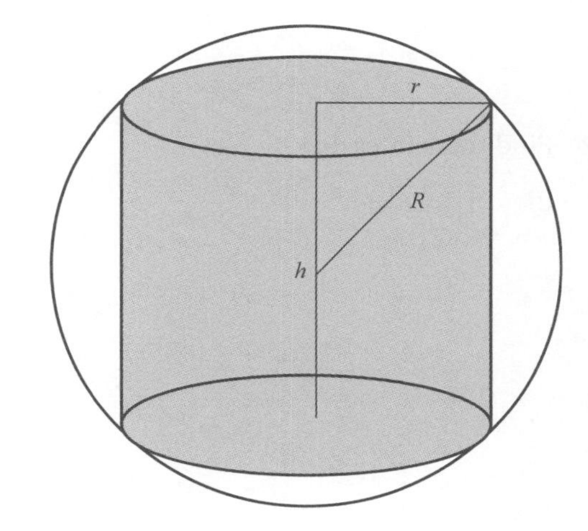

4 Three towns A, B and C form an isosceles triangle as shown in the diagram. They are to be joined by three roads: AD, DB and DC. Determine the position of the point D in the triangle so that the total length of the three roads is as small as possible (this point is known as the *Steiner Point*).

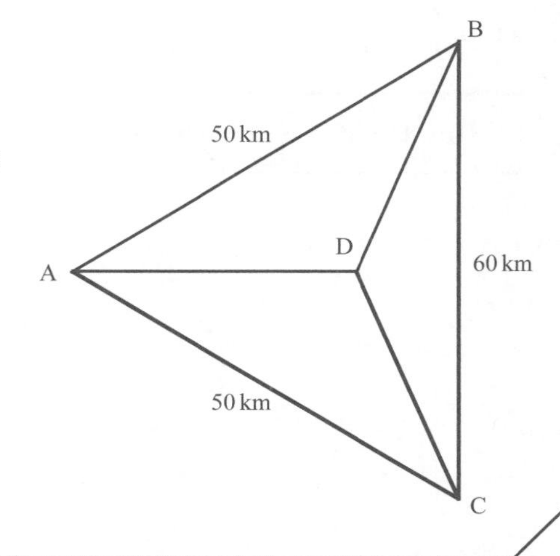

7 Integration

7.1 Reversing differentiation

1 Find the following integrals.

(i) $\int (2x^3 + x)\,dx$

(ii) $\int (1 + 3x)\,dx$

(iii) $\int \frac{4}{x^3}\,dx$

(iv) $\int 6\sqrt[3]{x}\,dx$

2 Find these integrals.

(i) $\int \left(\frac{2}{\sqrt{x}}\right) dx$

(ii) $\int 10\sqrt{x^3}\,dx$

3 Find $\int \frac{5x^2 - 1}{x^4}\,dx$.

Hint: Divide first.

4 A curve is such that $\frac{dy}{dx} = \frac{3}{x^4} + 1$ and A(−1, 2) is a point on the curve. Find the equation of the curve.

5 Find f(x) if $f'(x) = 3\sqrt{x} - 2x$ and f(4) = 30.

6 A curve is such that $\frac{dy}{dx} = 1 - \frac{k}{x^2}$, where k is a constant. The points M(1, 2) and N(−3, −6) are two points on the curve. Find the equation of the curve.

7.2 Finding areas

1 Find the area between the curve $y = x^2 - 4x + 3$ and the x-axis.

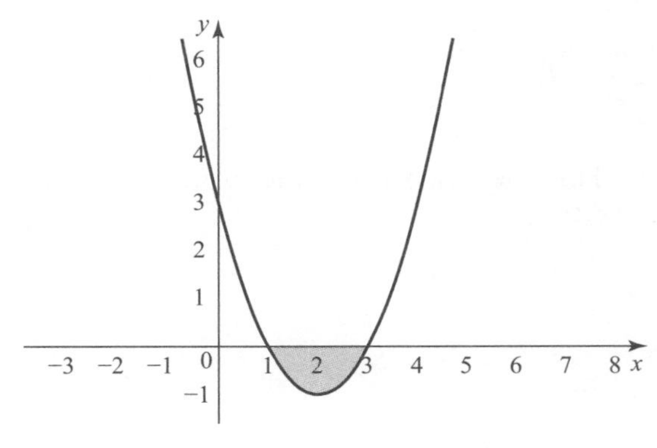

2 Find the area between the curve $y = 2\sqrt{x}$, the y-axis and the line $y = 4$.

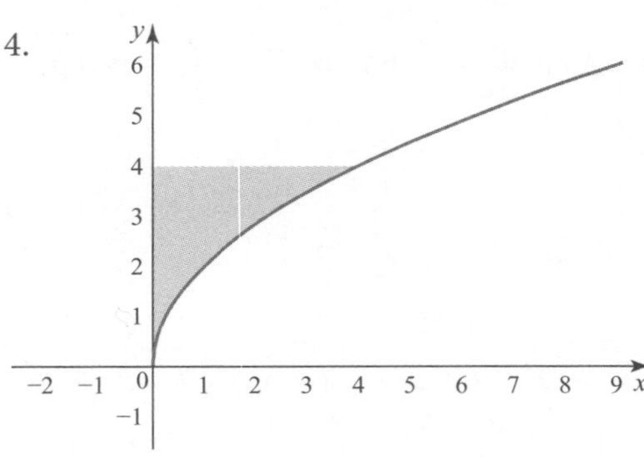

3 Find the area between the x-axis, the curve $f(x) = 3x^2 + 1$ and the lines $x = -1$ and $x = 2$.

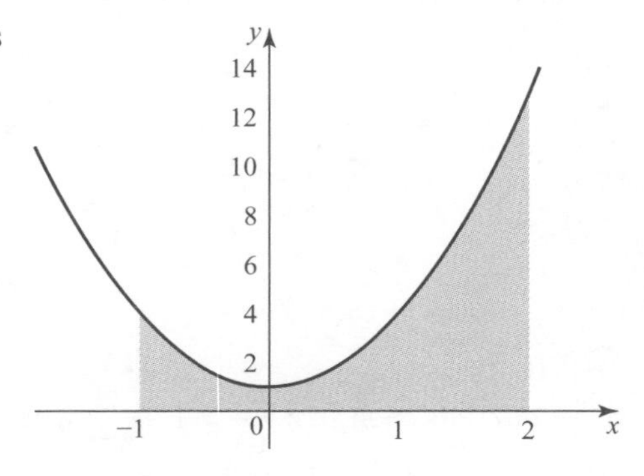

4 Find the area between the line $y = x - 1$ and the curve $y = x^2 - 2x - 1$.

> **Hint**: It does not matter that one curve is below the x-axis; the formula still works.

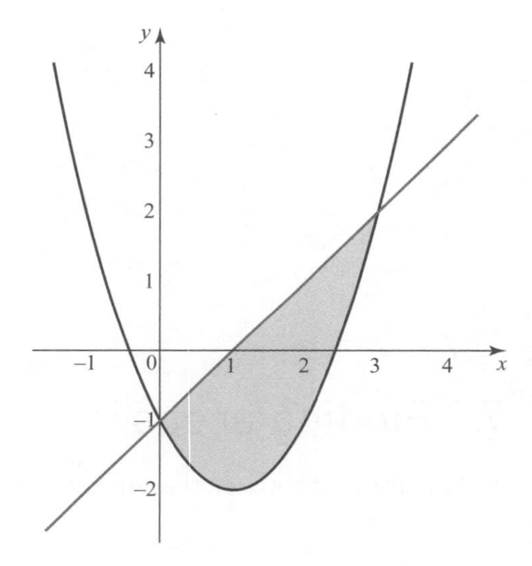

5 Find the area between the curves $y = 9 - x^2$ and $y = x^2 - 2x - 3$. Start by drawing a sketch of the two graphs.

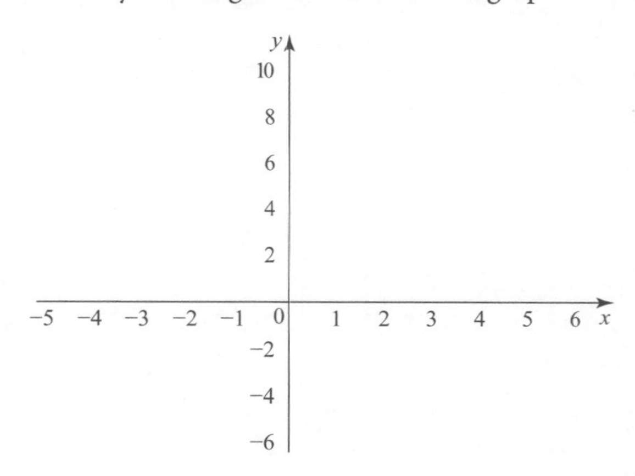

6 Find the area of the region between the curve $y = 2x^2 + 1$, the y-axis and the lines $y = 3$ and $y = 9$.

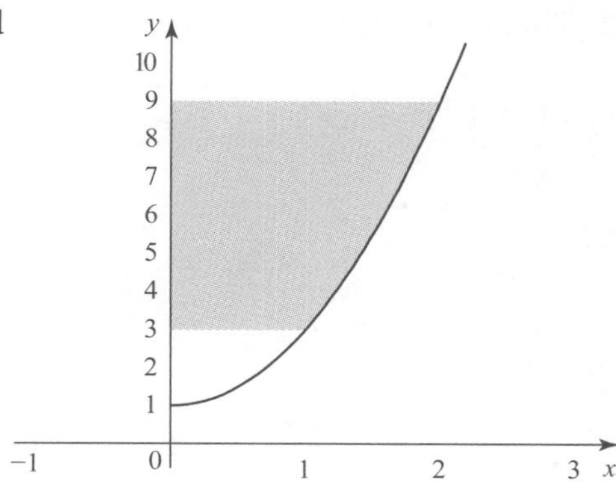

7 The area between the curve $y = 3\sqrt{x}$, the x-axis and the lines $x = 1$ and $x = k$ is 14.

Find the value of k.

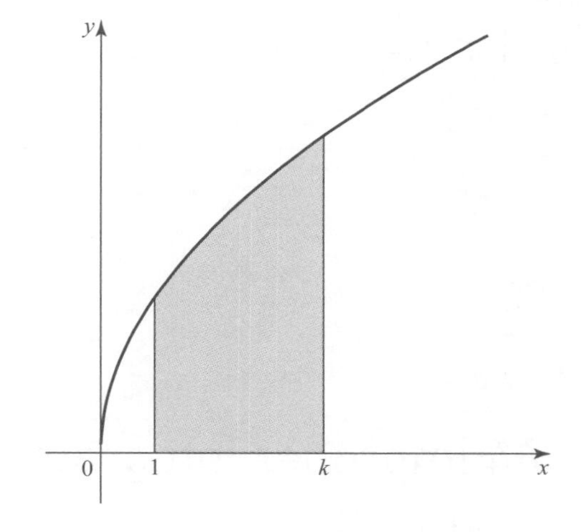

7.3 The reverse chain rule

1 Find these integrals.

(i) $\int (x-2)^3 \, dx$

(ii) $\int 2(1-6x)^9 \, dx$

(iii) $\int \left(\frac{x}{4}+3\right)^3 dx$

(iv) $\int (2x-1)^{\frac{1}{2}} \, dx$

2 Find these integrals.

(i) $\int \sqrt{1-x}\,\mathrm{d}x$

(ii) $\int \frac{2}{3(x-4)^5}\,\mathrm{d}x$

(iii) $\int \frac{3}{\sqrt{5-2x}}\,\mathrm{d}x$

(iv) $\int \frac{12}{\sqrt[3]{1+\frac{x}{2}}}\,\mathrm{d}x$

7.4 Improper integrals

1 Find $\int_0^1 \frac{1}{(2x+1)^3}\,\mathrm{d}x$.

2 Evaluate $\int_1^2 \frac{2}{\sqrt{x-1}}\,\mathrm{d}x$.

3 Evaluate $\int_0^\infty \frac{3}{(1+x)^2}\,\mathrm{d}x$.

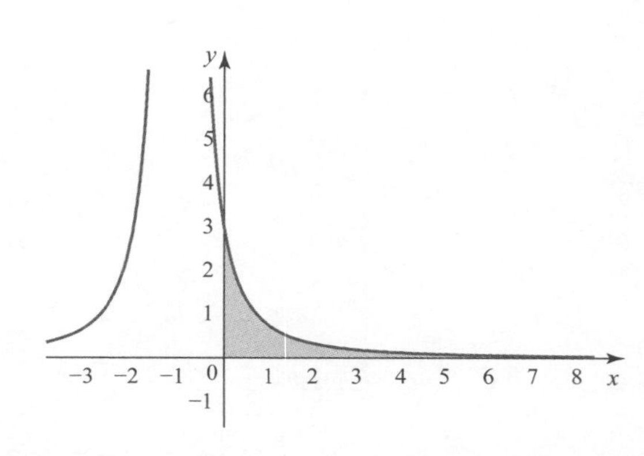

7.5 Finding volumes by integration

1 **(i)** Find the volume of the solid formed when the area between the curve $y = \dfrac{4}{x}$ and the x-axis for $x \geqslant 2$ is rotated through 360° about the x-axis. Give your answer in terms of π and to 3 significant figures.

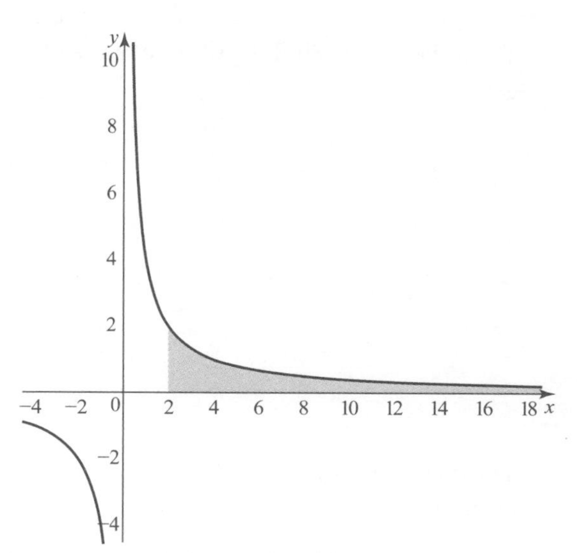

(ii) Find the volume of the solid formed when the area between the curve $y = \dfrac{4}{x}$, the x-axis and the lines $x = 3$ and $x = 4$ is rotated completely around the x-axis. Give your answer in terms of π and to 3 significant figures.

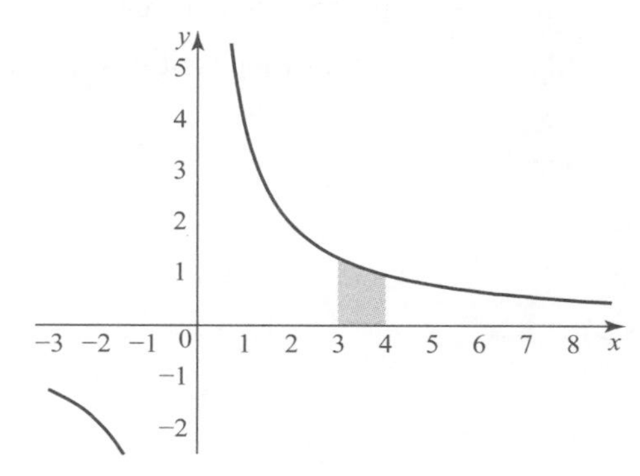

2 Find the volume of the solid formed when the shaded area under the curve $y = \dfrac{3}{2 - x}$ is rotated completely around the x-axis. Give your answer in terms of π.

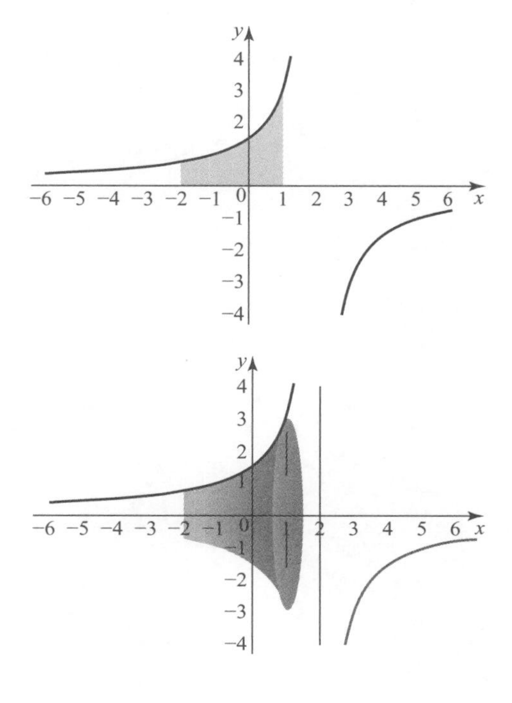

3 The part of the curve $y = 2(x^2 - 1)$ between $x = 1, x = 2$ and the x-axis is rotated completely about the y-axis. Find the volume of the solid generated.

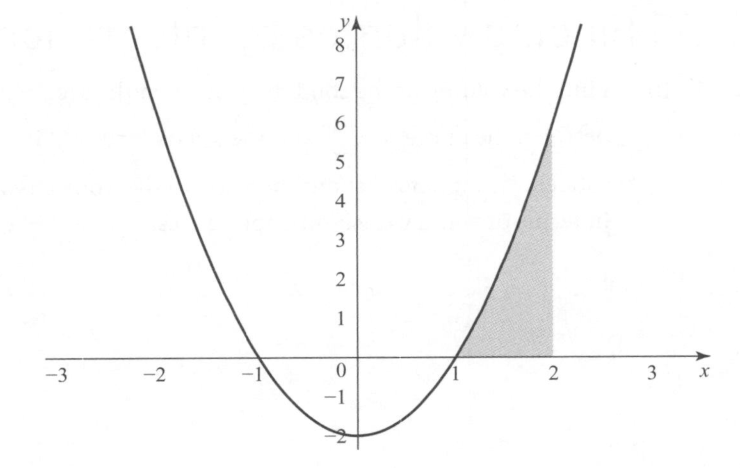

4 The diagram shows part of the curve $y = 4x + \frac{4}{x}$, which has a minimum point at M.

The line $y = 10$ intersects the curve at the points A and B.

(i) Find the coordinates of A, B and M.

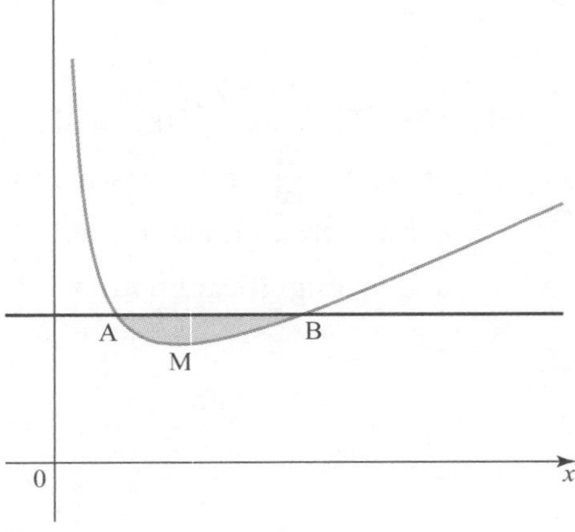

(ii) Find the volume obtained when the shaded region is rotated through 360° about the x-axis.

Further practice

1 Find these indefinite integrals.

(i) $\int(2x-5x^4)\mathrm{d}x$ **(ii)** $\int\left(\frac{x}{2}+2x^5\right)\mathrm{d}x$ **(iii)** $\int\left(\frac{1}{2x^2}+6x\right)\mathrm{d}x$

(iv) $\int x(2x-3\sqrt{x})\mathrm{d}x$ **(v)** $\int(3x+1)^5\,\mathrm{d}x$ **(vi)** $\int\frac{4}{(2x+1)^3}\mathrm{d}x$

2 Find the equation of the curve through A(1, –2) for which $\frac{\mathrm{d}y}{\mathrm{d}x}=2x+1$.

3 Find the area between the line $y=2$ and the curve $y=6-x^2$.

4 Find the area between the curve $y=\frac{2}{\sqrt[3]{x}}-2$, the x-axis and the lines $x=0$ and $x=8$.

5 Find the area between the curve $y=2\sqrt{x}+1$, the y-axis and the lines $y=3$ and $y=5$.

6 Find the area between the curves $y=4\sqrt{x}$ and $y=x$.

7 Evaluate **(i)** $\int_0^1 x^{-\frac{1}{2}}\,\mathrm{d}x$ **(ii)** $\int_1^\infty x^{-2}\,\mathrm{d}x$.

8 The area between the line $y=3x$, the x-axis and the lines $x=0$ and $x=2$ is rotated through 360° about the x-axis. Find the volume of the solid generated.

Check that your answer is correct by finding the volume using the formula for the volume of a cone, $V=\frac{1}{3}\pi r^2h$.

9 Find the volume of the solid formed when the area under the curve $y=\frac{3}{2x+1}$ between $x=0$ and $x=2$ is rotated completely around the x-axis. Give your answer in terms of π.

10 The diagram shows the line $y=1-x$ and the curve $y=\sqrt{1-x}$, which intersect at $(0,1)$ and $(1,0)$.

(i) Find the area of the shaded region.

(ii) Find the volume obtained when the shaded region is rotated through 360° about

(a) the x-axis

(b) the y-axis.

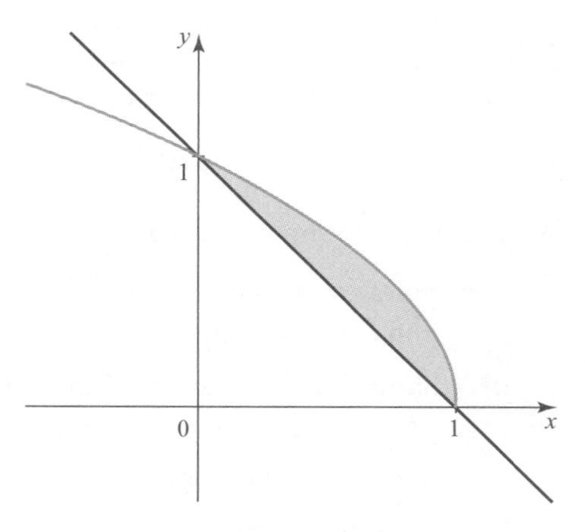

11 Find the volume of the solid formed when the area between the curve $y=x^3$ and the line $y=x$ is rotated through 360° about

(i) the x-axis

(ii) the y-axis.

Past exam questions

1 The diagram shows the straight line $x+y=5$ intersecting the curve $y=\frac{4}{x}$ at the points A(1, 4) and B(4, 1).

Find, showing all necessary working, the volume obtained when the shaded region is rotated through 360° about the x-axis. [7]

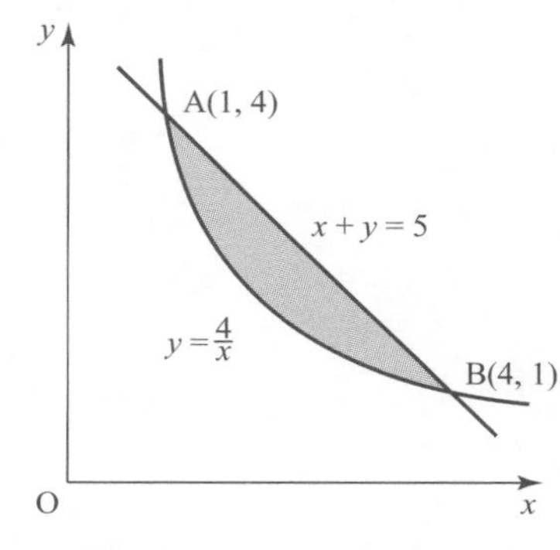

Cambridge International AS & A Level Mathematics 9709 Paper 12 Q6 June 2017

2 A curve is such that $\frac{\mathrm{d}y}{\mathrm{d}x} = \frac{8}{(5-2x)^2}$. Given that the curve passes through (2, 7), find the equation of the curve. [4]

Cambridge International AS & A Level Mathematics 9709 Paper 12 Q2 June 2016

3 The diagram shows parts of the curves $y = (2x-1)^2$ and $y^2 = 1 - 2x$, intersecting at points A and B.

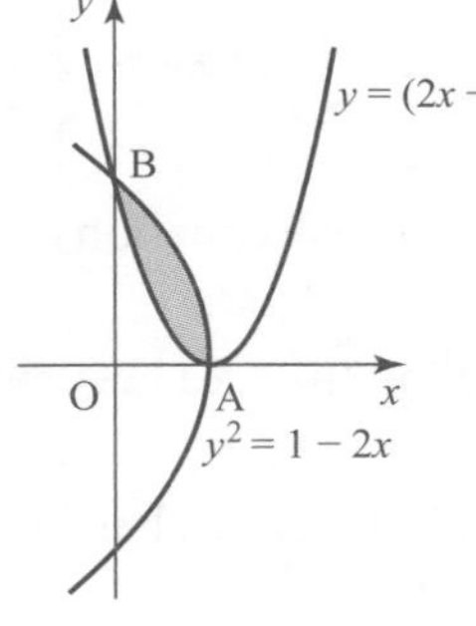

(i) State the coordinates of A. [1]

(ii) Find, showing all necessary working, the area of the shaded region. [6]

Cambridge International AS & A Level Mathematics 9709 Paper 11 Q7 November 2016

4 A curve is such that $\frac{\mathrm{d}y}{\mathrm{d}x} = \frac{12}{\sqrt{4x+a}}$, where a is a constant. The point P(2, 14) lies on the curve and the normal to the curve at P is $3y + x = 5$.

(i) Show that $a = 8$. [3]

(ii) Find the equation of the curve. [4]

Cambridge International AS & A Level Mathematics 9709 Paper 13 Q6 June 2014

5 A curve is such that $\frac{\mathrm{d}^2y}{\mathrm{d}x^2} = \frac{24}{x^3} - 4$. The curve has a stationary point at P where $x = 2$.

(i) State, with a reason, the nature of this stationary point. [1]

(ii) Find an expression for $\frac{\mathrm{d}y}{\mathrm{d}x}$. [4]

(iii) Given that the curve passes through the point (1, 13), find the coordinates of the stationary point P. [4]

Cambridge International AS & A Level Mathematics 9709 Paper 12 Q10 November 2014

6 The diagram shows parts of the curves $y = 9 - x^3$ and $y = \frac{8}{x^3}$ and their points of intersection P and Q. The x-coordinates of P and Q are a and b respectively.

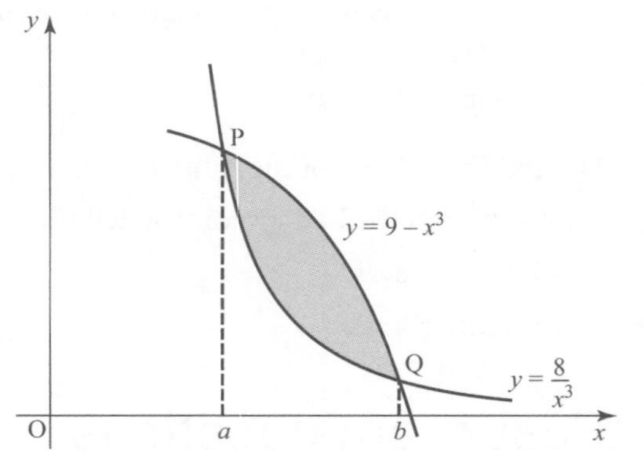

(i) Show that $x = a$ and $x = b$ are roots of the equation $x^6 - 9x^3 + 8 = 0$. Solve this equation and hence state the value of a and the value of b. [4]

(ii) Find the area of the shaded region between the two curves. [5]

(iii) The tangents to the two curves at $x = c$ (where $a < c < b$) are parallel to each other. Find the value of c. [4]

Cambridge International AS & A Level Mathematics 9709 Paper 13 Q11 November 2010

STRETCH AND CHALLENGE

1 The shaded region between the curve $y = (x + 2)^2$ and the line $y = 2x + k$ between $x = -2$ and $x = 0$ has an area of $\frac{10}{3}$ square units.

Find the value of the constant k.

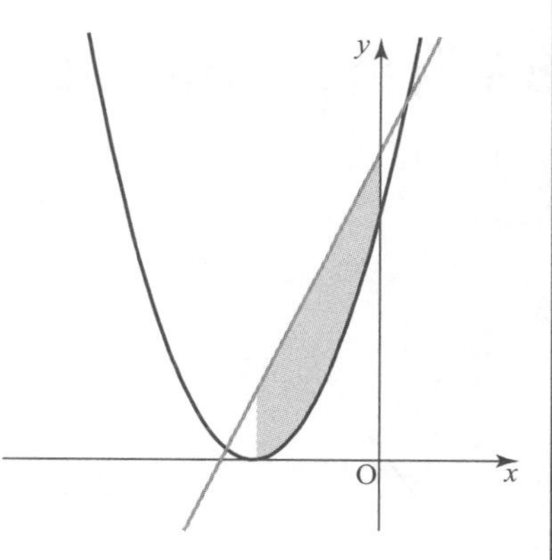

2 A symmetrical plant pot, shown here, has a circular base of radius r centimetres, a circular top of radius R centimetres, and straight sloping sides.
Show by integration that the volume of the pot is given by $\frac{1}{3}\pi h\left(R^2 + rR + r^2\right)$ cm^3.

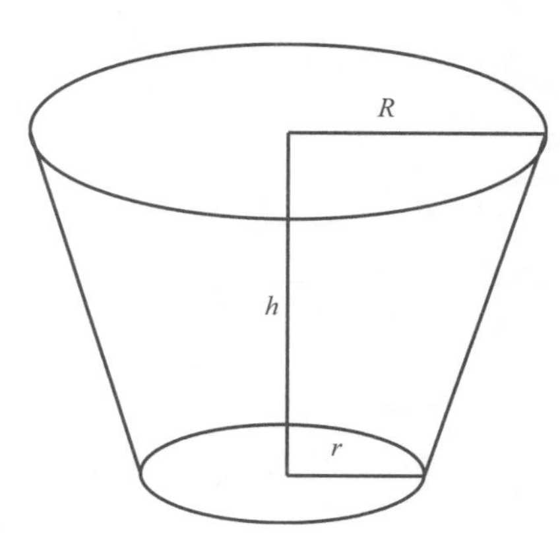

3 The region under the function $f(x) = 2x^2, x > 0$, between $x = k$ and $x = k + 1$, has an area of $\frac{109}{9}$ square units. Find the value of k.

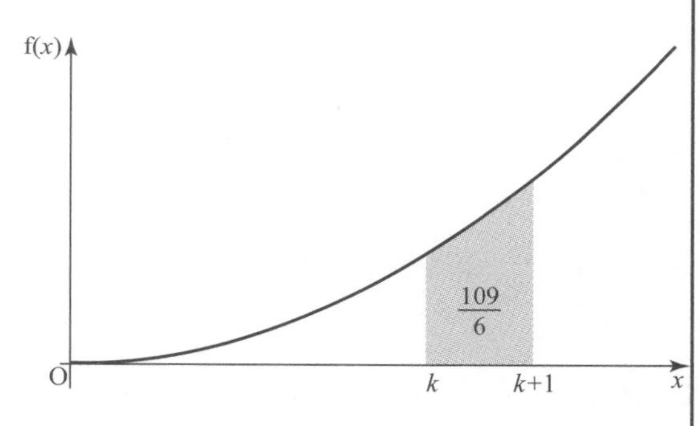

4 A ring-maker has a collection of solid gold rings for different-sized fingers. The cross section of each ring is a segment of a circle radius R as shown in the diagram. All rings in the collection have the same width w.
The ring-maker says that, although they have different diameters, the rings all contain the same amount of gold.
Is this true? Justify your answer.

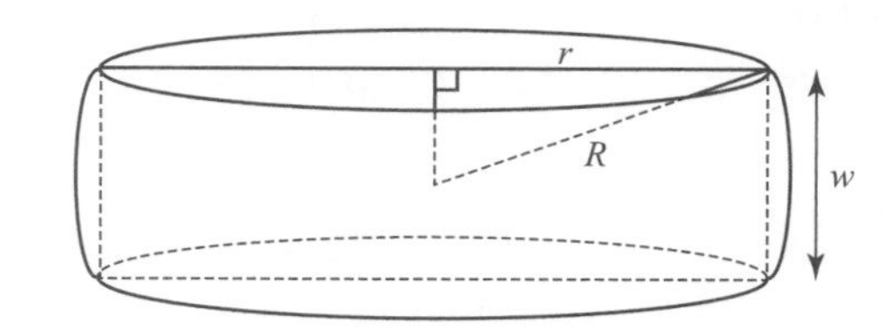

Hint: The equation of a circle with radius R is $x^2 + y^2 = R^2$.

8 Trigonometry

8.1 Trigonometrical functions

1 Find the lengths and angles marked.

(i)

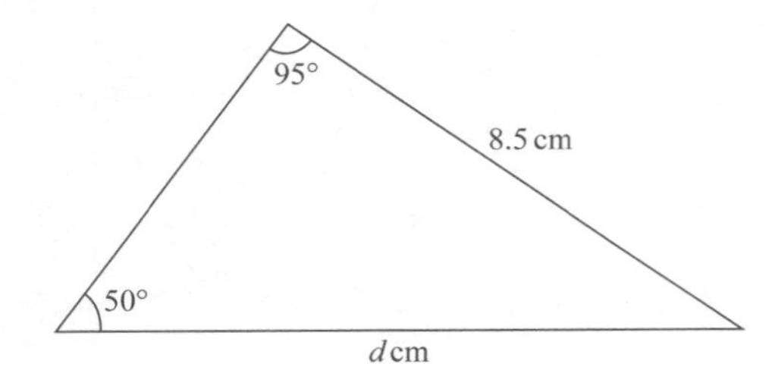

(ii)

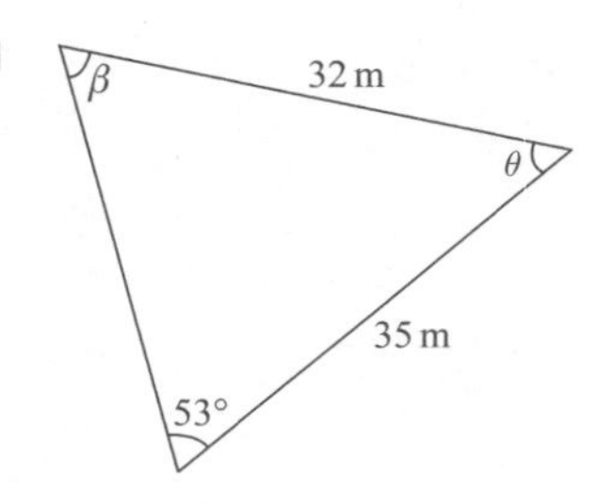

(iii)

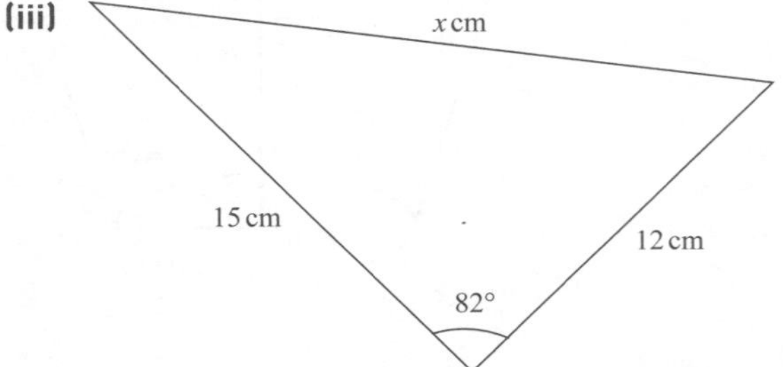

(iv)

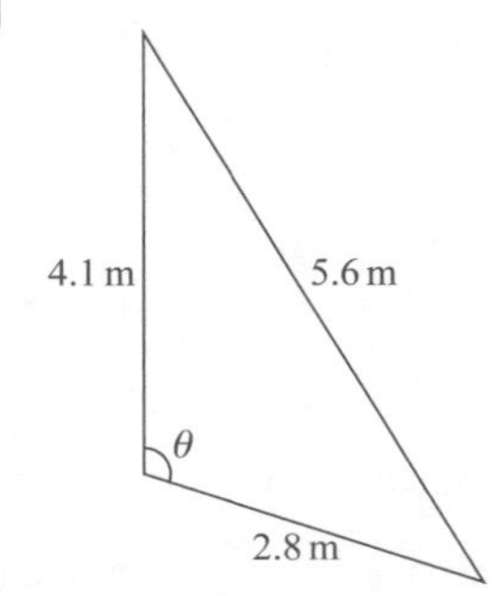

2 Find the areas of the triangles in question 1.

(i)

(ii)

(iii)

(iv)

3 Find the unknown lengths marked in the following triangles.

(i)

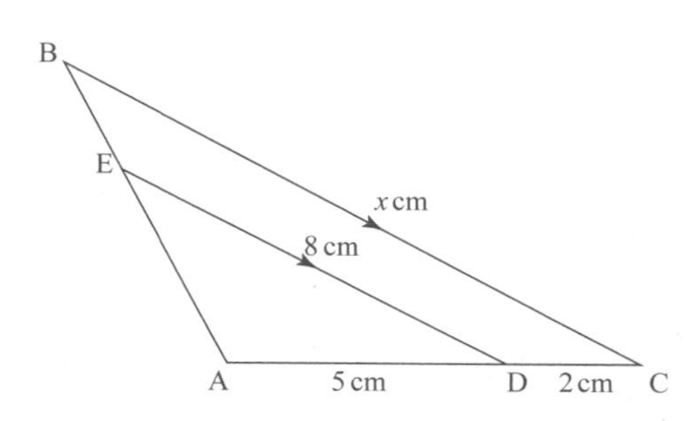

(ii)

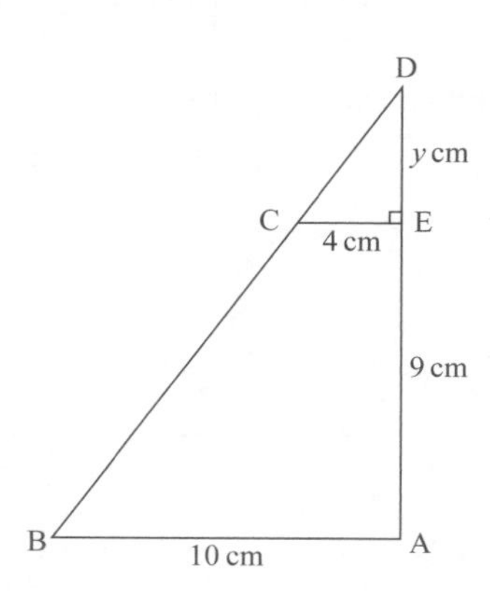

8.2 Trigonometrical functions for angles of any size

1 Use the unit circle to find the missing ratios and exact values. Each angle is between 0° and 360°.

First ratio	Other ratio that has the same value	Exact value
sin 30°	sin ____°	
sin 210°	sin ____°	
cos ____°	cos ____°	$\frac{1}{2}$
tan 30°	tan ____°	
cos 150°	cos ____°	
tan ____°	tan ____°	$-\sqrt{3}$

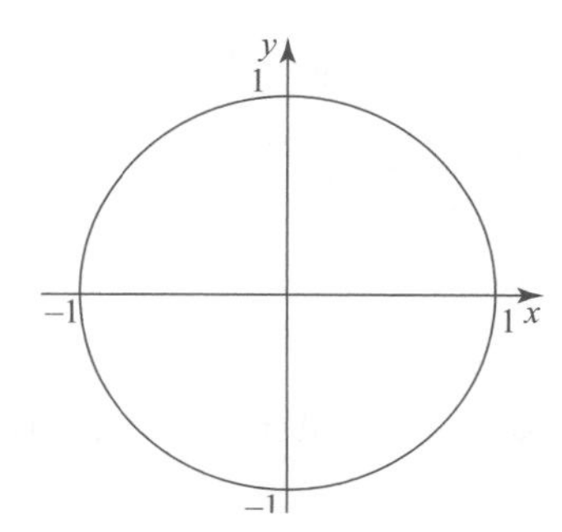

2 Given that $x = \cos^{-1}\left(\frac{1}{4}\right)$ and x is an acute angle, find the exact value of these.

(i) $\sin x$

(ii) $\tan^2 x$

3 The acute angle θ radians is such that $\cos\theta = k$, where k is a positive constant and $0 \leqslant \theta \leqslant \frac{\pi}{2}$.
Express the following in terms of k.

(i) $\cos(-\theta) \quad =$

(ii) $\cos(\pi - \theta) \quad =$

(iii) $\sin\theta \quad =$

(iv) $\cos(\pi + \theta) \quad =$

(v) $\tan\theta \quad =$

8.3 Trigonometrical graphs

1 Sketch these graphs.

(i) $y = \sin 2x + 1 \quad 0° \leqslant x \leqslant 360°$

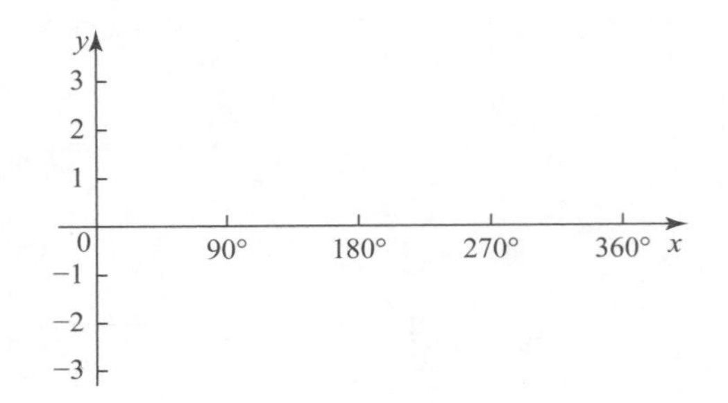

(ii) $y = 3\cos x - 2 \quad 0 \leqslant x \leqslant 2\pi$

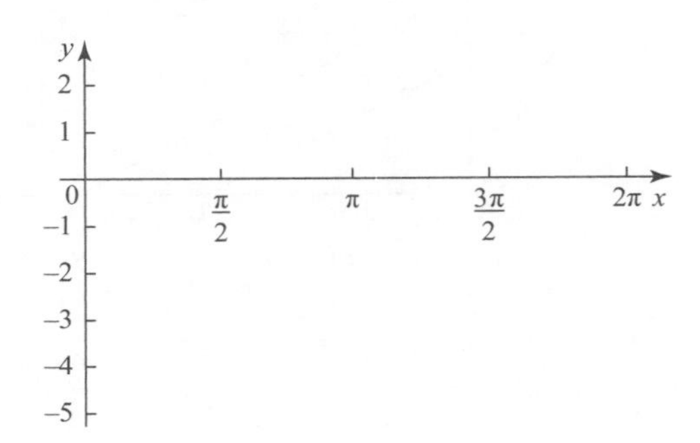

(iii) $y = \tan 2x \quad 0° \leqslant x \leqslant 360°$

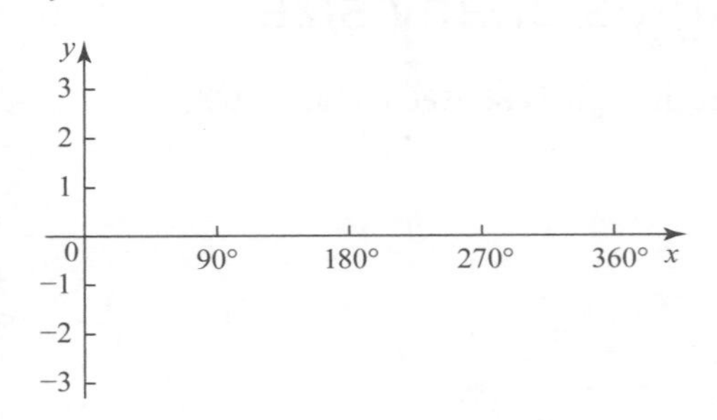

(iv) $y = 4 - 3\sin 2x \quad 0 \leqslant x \leqslant 2\pi$

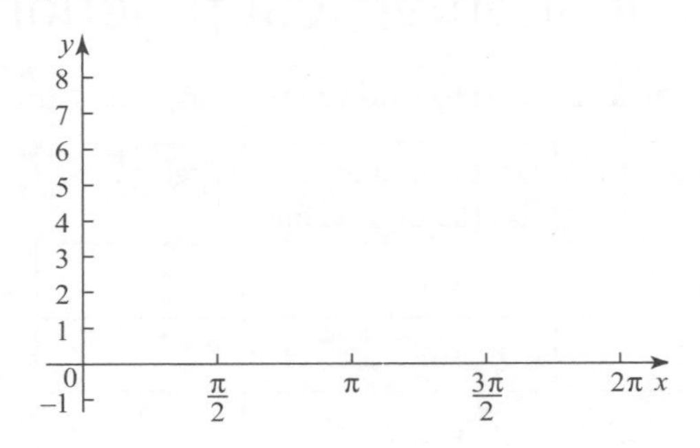

2 The diagram shows the graph of $y = A\sin Bx + C$. Write down the values of A, B and C.

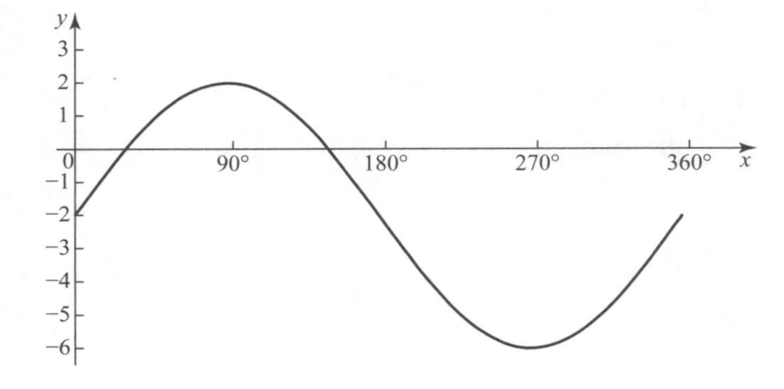

$A =$ __________

$B =$ __________

$C =$ __________

3 The diagram shows the graph of $y = A\cos Bx + C$. Write down the values of A, B and C.

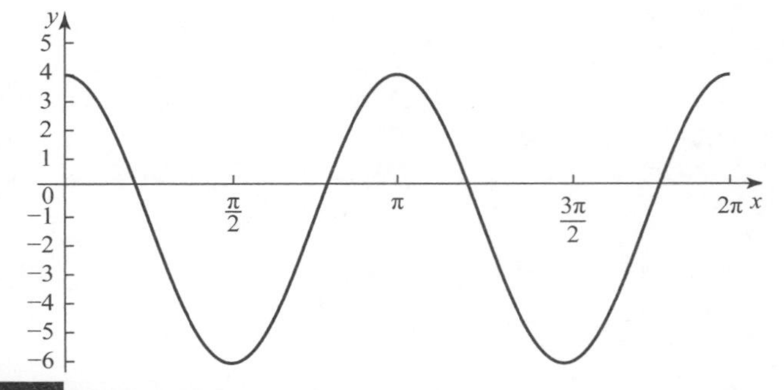

$A =$ __________

$B =$ __________

$C =$ __________

4 The diagram shows the graph of $y = A\sin Bx + C$. Write down the values of A, B and C.

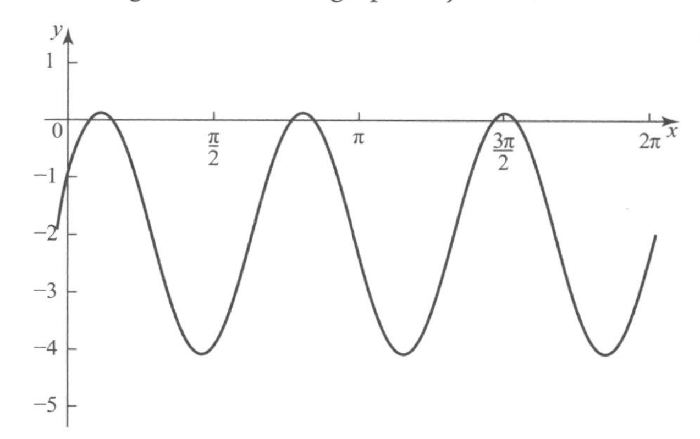

$A =$ ________

$B =$ ________

$C =$ ________

5 The diagram shows the graph of $y = \tan Ax + B$. Write down the values of A and B.

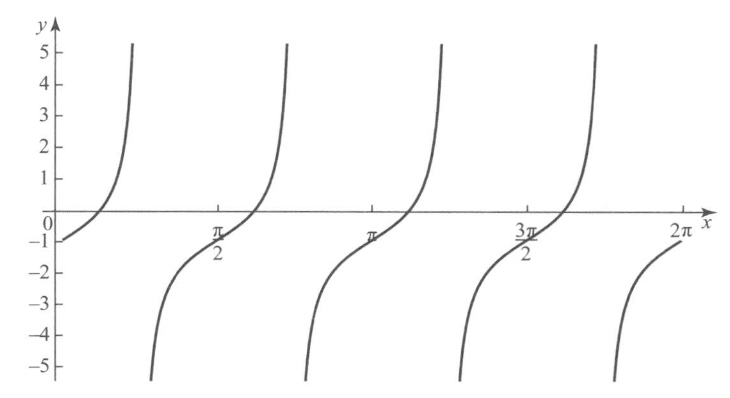

$A =$ ________

$B =$ ________

6 The temperature (T) of the water in a lake alternates regularly throughout the day. At t hours after 12 a.m. the temperature is given by the equation $T = 2\cos(3\pi t) - 1$.

(i) Find the period of T.

(ii) Sketch the graph of T between 12 a.m. and 4 a.m.

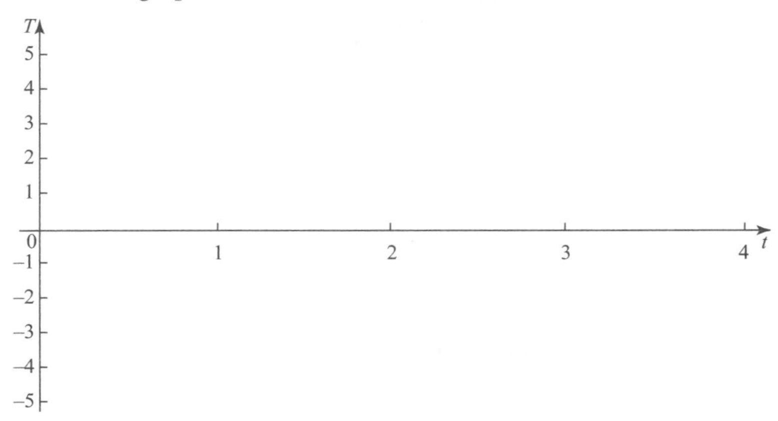

(iii) From the graph, find the warmest temperature of the water, and the times it reaches this temperature from 12 a.m. to 4 a.m.

7 The function f is defined by $f : x \mapsto a - b\sin x$, where a and b are both positive constants.

(i) The minimum value of f is -2 and the maximum value is 8. Find the values of a and b.

(ii) Hence sketch $y = f(x)$ for $0 \leqslant x \leqslant 2\pi$.

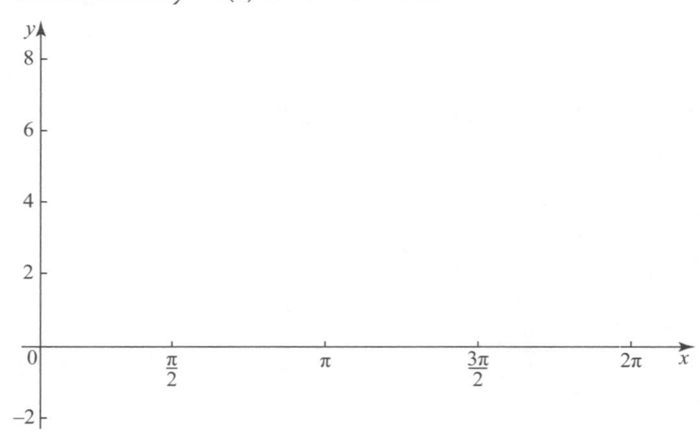

8.4 Identities

1 Prove the following identities.

(i) $\sin x \tan x \equiv \dfrac{1}{\cos x} - \cos x$

(ii) $\dfrac{1 + \cos x}{\sin x} \equiv \dfrac{\sin x}{1 - \cos x}$

(iii) $\tan^2 x - \sin^2 x \equiv \tan^2 x \sin^2 x$

(iv) $\tan x + \dfrac{1}{\cos x} \equiv \dfrac{\sin x}{1 - \sin x}$

2 Prove the following identities.

(i) $\dfrac{1}{1 - \cos x} + \dfrac{1}{1 + \cos x} \equiv \dfrac{2}{\sin^2 x}$

(ii) $\dfrac{\cos x}{1 - \cos x} - \dfrac{\cos x}{1 + \cos x} \equiv \dfrac{2}{\tan^2 x}$

8.5 Trigonometrical equations

1 Solve the following equations where $0° \leqslant x \leqslant 360°$.

(i) $2\sin x = -1$

(ii) $\cos 2x = \frac{1}{2}$

2 Solve $2\cos 3x - \sin 3x = 0$ for $0° \leqslant x \leqslant 180°$.

3 Show, on the unit circles, the two angles between 0° and 360° that satisfy the given equations. Write down the angles.

(i) $\sin x = \frac{1}{4}$

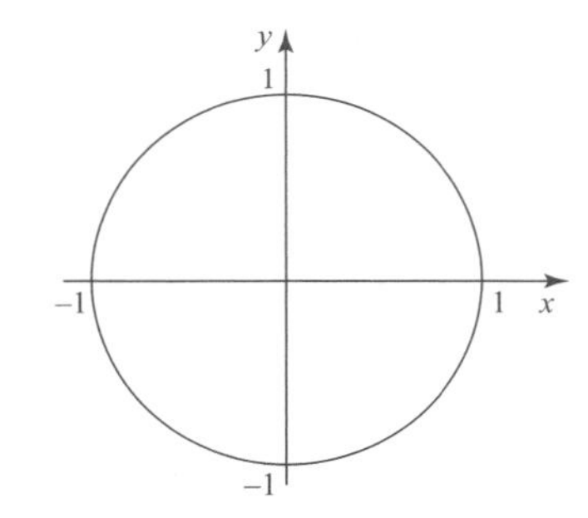

(ii) $\cos x = -0.7$

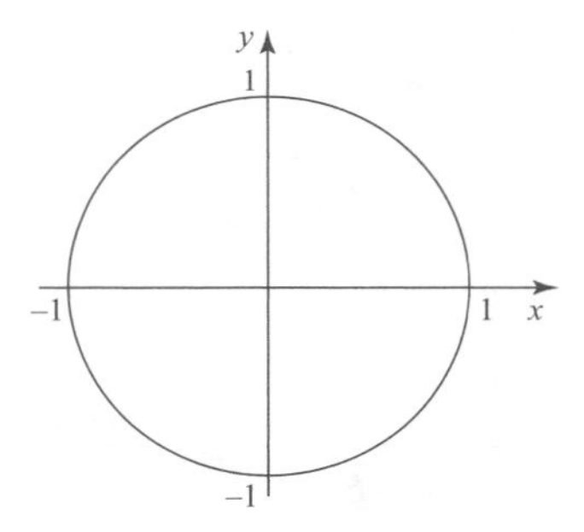

(iii) $\tan x = \frac{7}{8}$

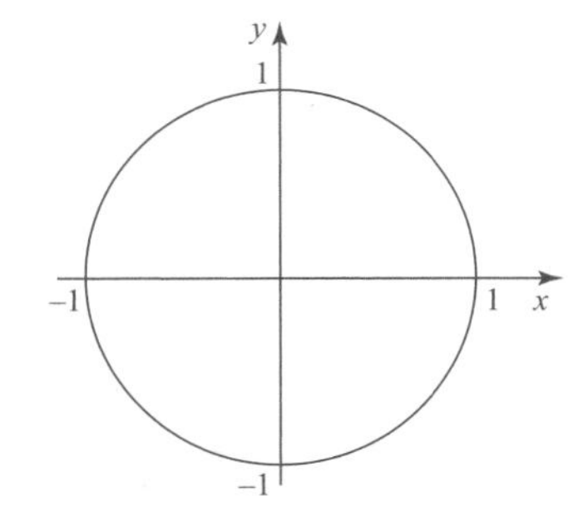

(iv) $\sin x = -\frac{2}{5}$

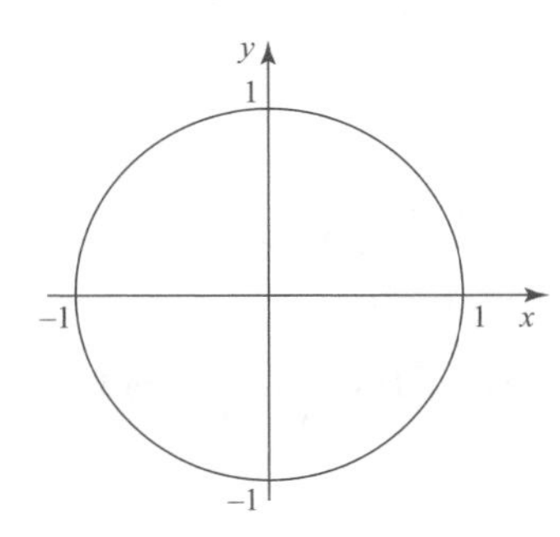

4 Solve the following equations for $0 \leqslant x \leqslant 2\pi$. (Give your answers in radians.)

(i) $2\sin x = -\sqrt{3}$

(ii) $\cos 3x = -1$

5 The height of a carriage above the ground (h) on a Ferris wheel ride after t seconds is given by the equation

$$h = 54 + 53\sin\left(\frac{\pi}{20}t\right)$$

(i) What is the period of the function (i.e. how long does one complete revolution of the wheel take)?

(ii) Find the maximum height of the ride.

(iii) Find how long it takes after the start of the ride to get to a height of 30 m off the ground.

6 **(i)** Prove the identity $\left(\frac{1}{\cos\theta} + \tan\theta\right)^2 \equiv \frac{1 + \sin\theta}{1 - \sin\theta}$.

(ii) Hence solve the equation $\left(\frac{1}{\cos\theta} + \tan\theta\right)^2 = \frac{3}{7}$ for $0° \leqslant \theta \leqslant 360°$.

7 Solve the following equations over the given intervals.
Use the unit circle diagram to help you to find all the solutions.

(i) $2\cos^2 x = \cos x$ for $0° \leqslant \theta \leqslant 360°$

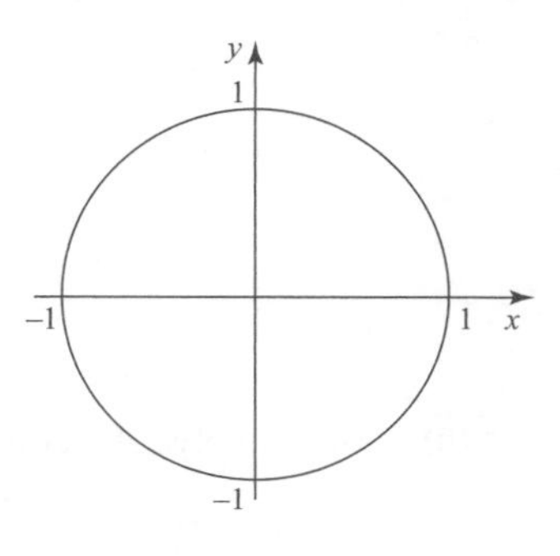

(ii) $2\tan x = \tan^2 x$ for $0 \leqslant x \leqslant 2\pi$

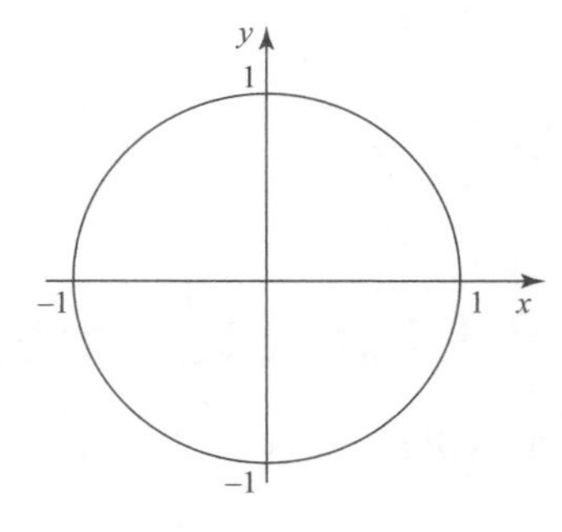

(iii) $\cos x = -3\tan x$ for $-\pi \leqslant x \leqslant \pi$

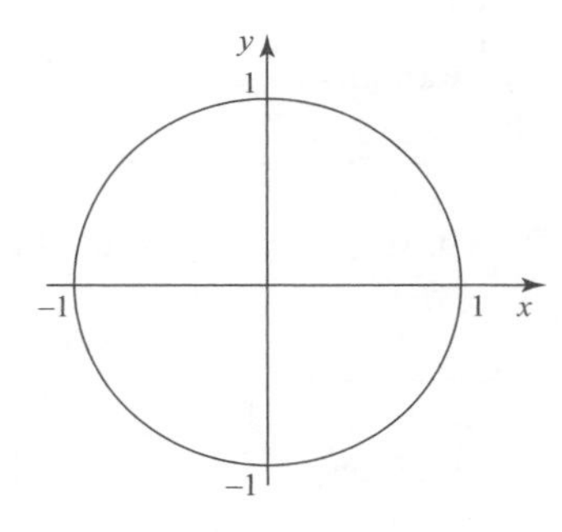

(iv) $4 + 5\cos x = 2\sin^2 x$ for $-180° \leqslant x \leqslant 180°$

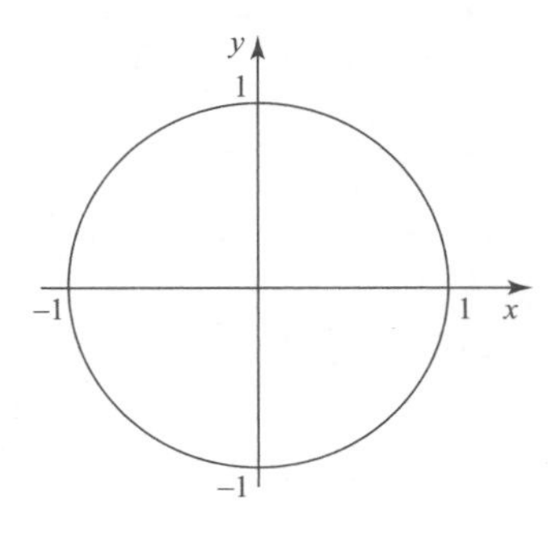

8 (i) Given that $4\cos^2 x + 7\sin x - 2 = 0$

show that, for real values of x, $\sin x = -\frac{1}{4}$

(ii) Hence solve the equation

$4\cos^2(\theta - 20°) + 7\sin(\theta - 20°) - 2 = 0$ for $0° \leqslant \theta \leqslant 360°$.

8.6 Circular measure

1 Fill in the spaces in the table. Give all your answers as exact fractions of π in the radians row.

Degrees	120°		330°		240°		225°	
Radians		$\frac{\pi}{5}$		$\frac{5\pi}{6}$		$\frac{3\pi}{10}$		$\frac{7\pi}{4}$

2 Fill in the spaces in the table. Give the answers in degrees to 1 decimal place and in radians to 3 significant figures.

Degrees	12°		145°		235°		342.5°	
Radians		0.190		2.95		5.04		1.37

3 Find the perimeter and area of the following shaded regions.

(i)

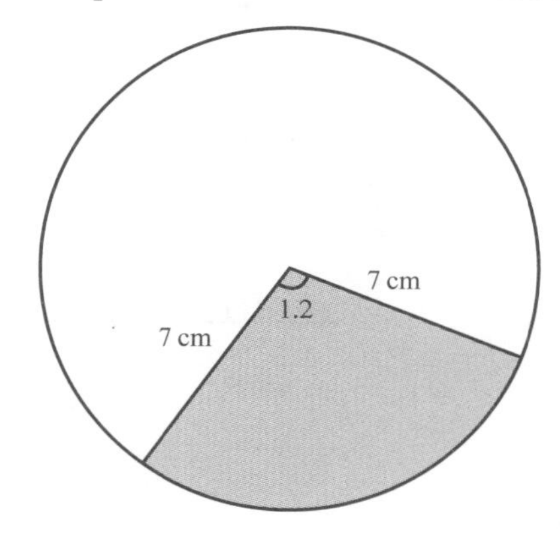

(ii)

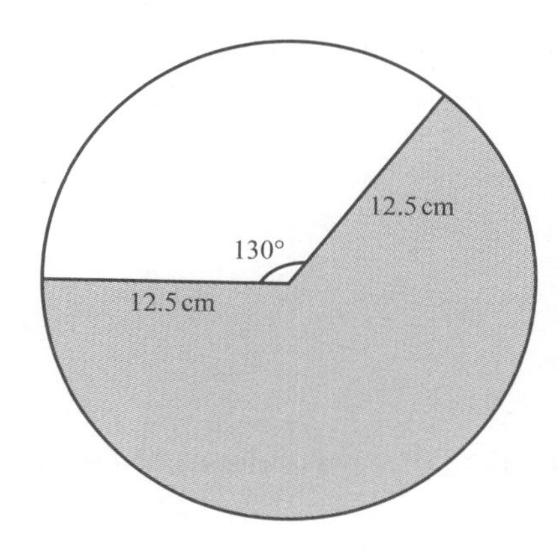

4 Find the value of θ in radians and degrees.

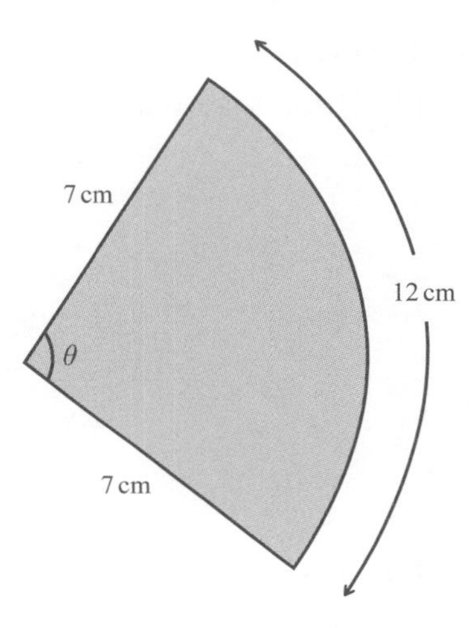

5 The diagram shows a circle centre O and radius 5 cm.

The tangents at A and B meet at the point C. Angle AOB is $\frac{2\pi}{3}$ radians.

Find the perimeter and area of the shaded region.

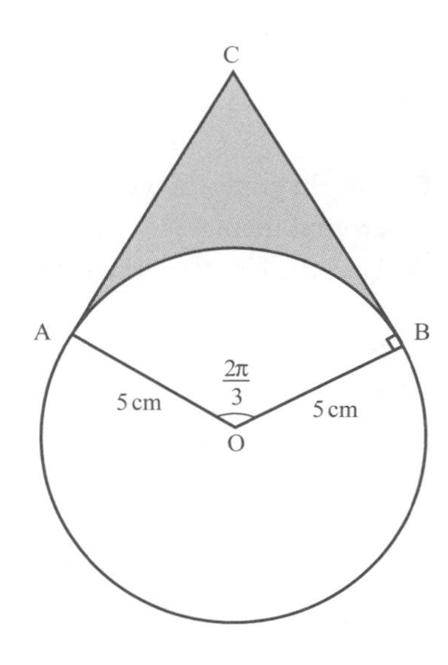

6 In the diagram, OQR is a sector of a circle with centre O and radius 10 cm. Angle QOR = 0.8 radians and RP is the perpendicular from R to OQ.

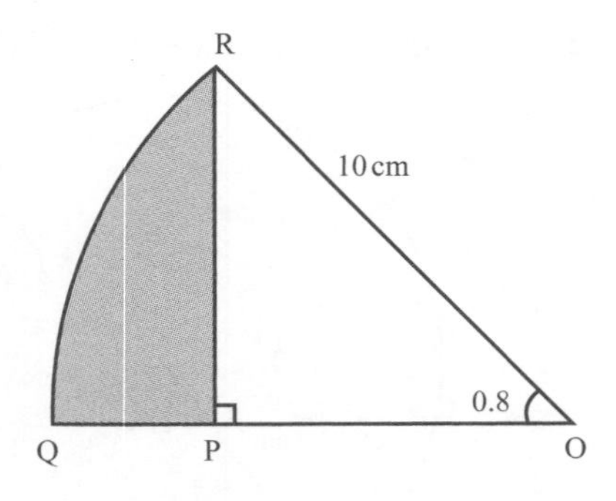

(i) Find the perimeter of the shaded region

(ii) Find the area of the shaded region.

7 The diagram shows a sector of a circle radius 6 cm.

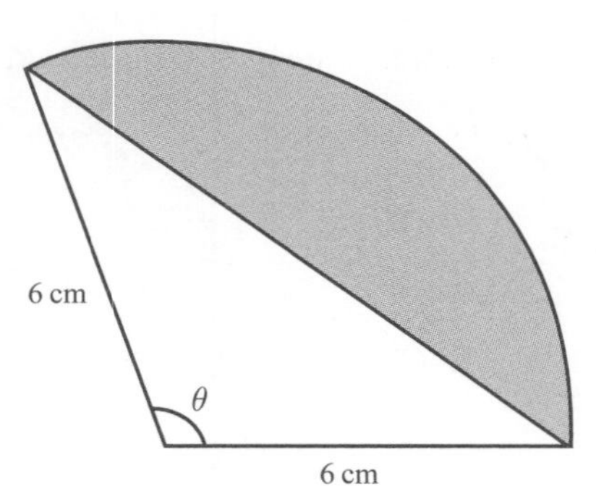

(i) If the shaded area is 36 cm^2, show that $\theta = \sin\theta + 2$.

(ii) Given that the perimeter of the sector is 27.3 cm, find the value of θ.

8 The diagram shows three circles of radius 4 cm with a rubber band fitted around them.
Find the length of the rubber band.

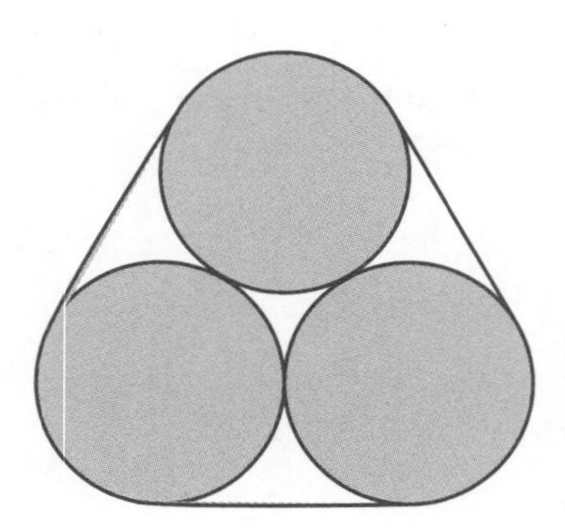

9 The diagram shows a circle with centre O and radius 8 cm.
The line AC is a tangent to the circle.

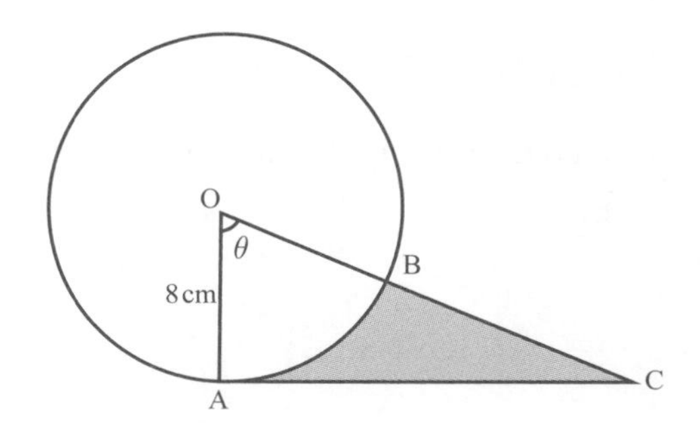

(i) Show that the area of the shaded region is $32(\tan\theta - \theta)\,\text{cm}^2$.

(ii) When $\theta = \frac{\pi}{3}$, find the perimeter of the shaded region.

10 The diagram shows a sector, OAB, of a circle, with centre O and radius 16 cm.
The midpoints of OB and OA are D and C, respectively. The length of DC is 8 cm.
AD is an arc of the circle with centre C and radius 8 cm. The shaded region is bounded by the line BD and the arcs AB and AD.

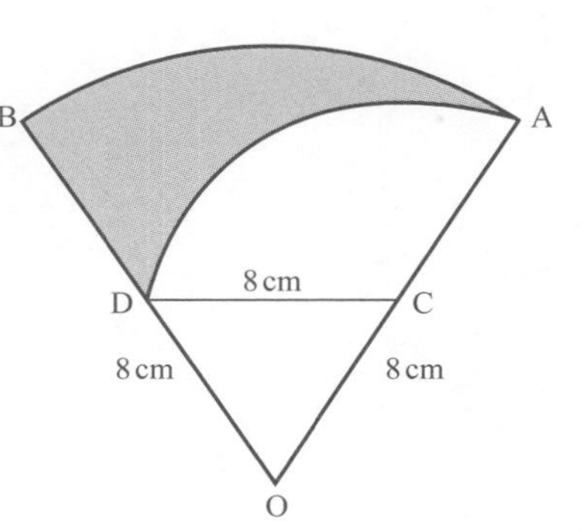

(i) Show that the angle ACD $= \frac{2}{3}\pi$ radians.

(ii) Show that the perimeter of the shaded region is $\left(\frac{32}{2}\pi + 8\right)$ cm.

(iii) Find the exact area of the shaded region.

Further practice

1 Find the value of the constants in the equations of the following graphs.

(i) $y = a \sin bx$

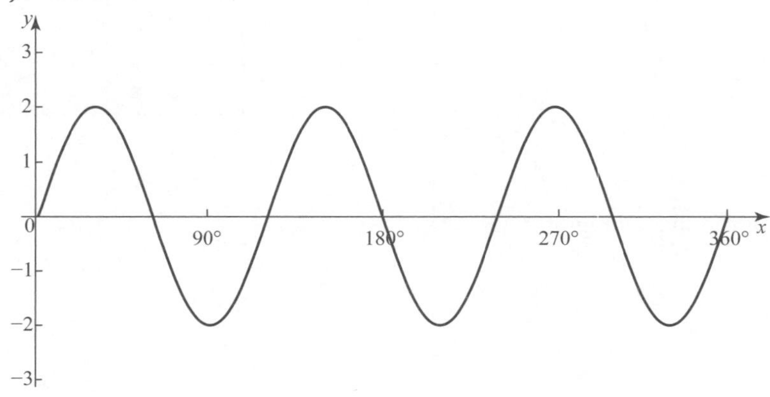

(ii) $y = a - b \cos cx$

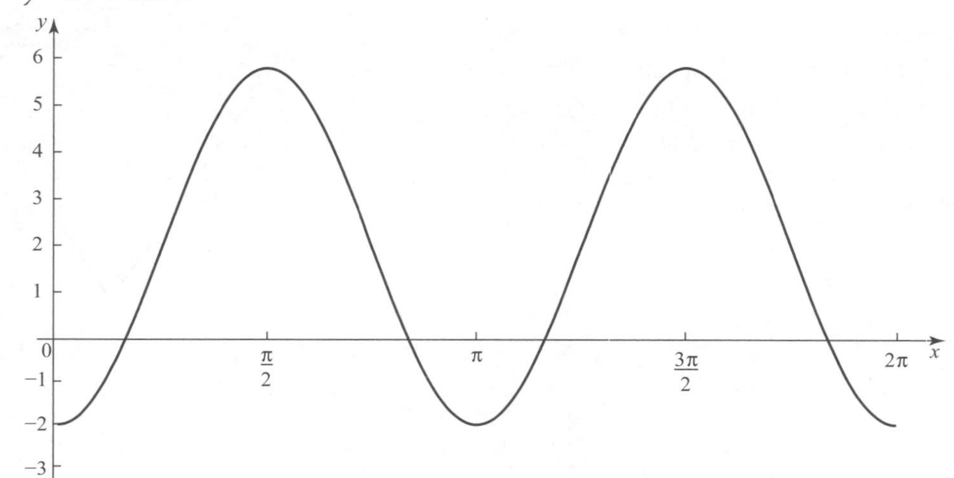

2 Prove these identities.

(i) $\tan x(1 - \sin^2 x) \equiv \sin x \cos x$ **(ii)** $\sin^4 x - \cos^4 x \equiv \sin^2 x - \cos^2 x$

(iii) $\dfrac{\cos\theta}{\tan\theta(1 + \sin\theta)} \equiv \dfrac{1}{\sin\theta} - 1$

3 Solve the following equations, where $0° \leqslant x \leqslant 360°$.

(i) $\tan x + 1 = 3$ **(ii)** $\sin(x - 30°) = 1$

4 Solve the following equations for $0 \leqslant x \leqslant 2\pi$. (Give your answers in radians.)

(i) $2\tan x - 1 = 1$ **(ii)** $\frac{1}{2}\sin(2x + 1) = 0.1$ **(iii)** $4\sin^2 x + 4\sin x - 3 = 0$

(iv) $2\sin^2 x = 3 - 3\cos x$ **(v)** $3\sin^2 x + 5\cos^2 x = 9\sin x$

5 The acute angle x radians is such that $\sin x = k$, where k is a positive constant. Express, in terms of k

(i) $\sin(\pi - x)$ **(ii)** $\cos x$ **(iii)** $\tan\left(\frac{\pi}{2} - x\right)$

6 Given that $\tan x = -\frac{2}{7}$ and $90° \leqslant x \leqslant 180°$, find the exact value of

(i) $\cos x$ **(ii)** $\sin^2 x$

7 Prove the identity $1 + \tan^2\theta \equiv \dfrac{1}{\cos^2\theta}$.

8 Prove the identity $\tan x + \dfrac{1}{\tan x} \equiv \dfrac{1}{\sin x \cos x}$.

9 Prove the identity $\dfrac{1 - \cos x}{\sin x} + \dfrac{\sin x}{1 - \cos x} \equiv \dfrac{2}{\sin x}$.

10 Solve these equations.

(i) $2\cos 2x = \sqrt{3}$ for $0° \leqslant x \leqslant 360°$

(ii) $\cos^2 x - 1 = \sin x$ for $-\pi \leqslant x \leqslant \pi$

(iii) $2\sin x \tan x = 3$ for $-\pi \leqslant x \leqslant \pi$

11 Find the area of the segment shown.

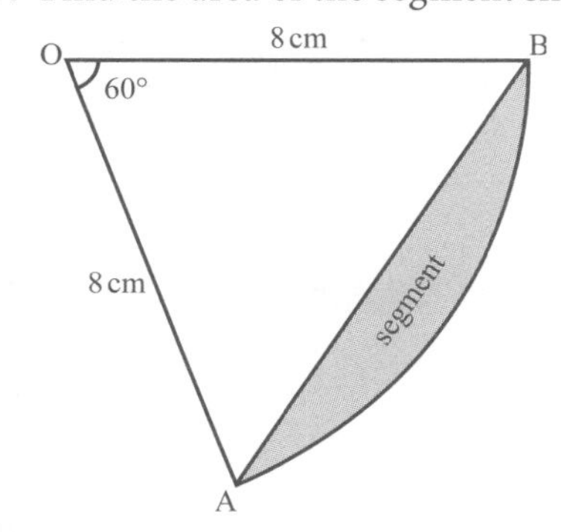

12 In the diagram, AC is an arc of a circle, centre O and radius 10 cm. The line AB is perpendicular to OA and OCB is a straight line. Angle AOC = $\frac{1}{3}\pi$ radians. Find the area and perimeter of the shaded region, giving your answer in terms of π and $\sqrt{3}$.

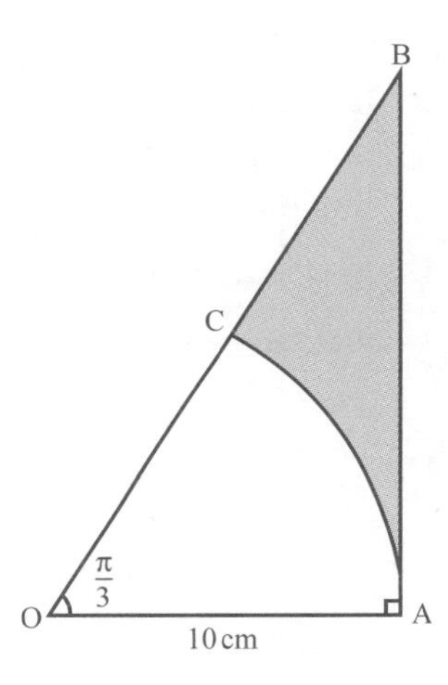

13 In the diagram, the circle has centre O and radius 6 cm.
The points D and F lie on the circle, and the arc length DF is 10 cm.
The tangents to the circle at D and F meet at the point E.

Calculate

(i) angle DOF in radians

(ii) the length of DE

(iii) the area of the shaded region.

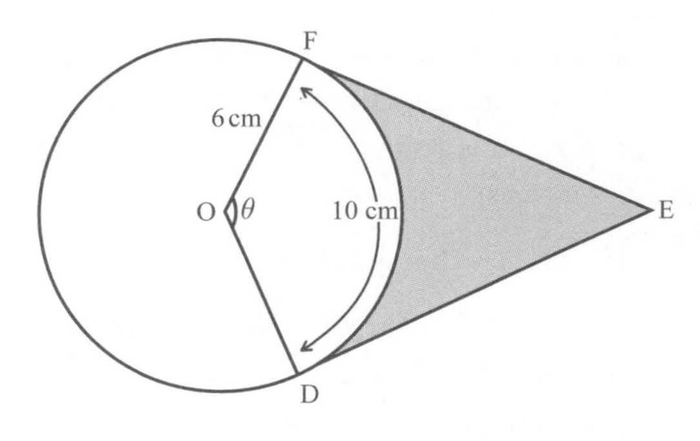

Past exam questions

1 The diagram shows a circle with radius r cm and centre O. Points A and B lie on the circle and ABCD is a rectangle. Angle AOB = 2θ radians and AD = r cm.

(i) Express the perimeter of the shaded region in terms of r and θ. [3]

(ii) In the case where $r = 5$ and $\theta = \frac{1}{6}\pi$, find the area of the shaded region. [4]

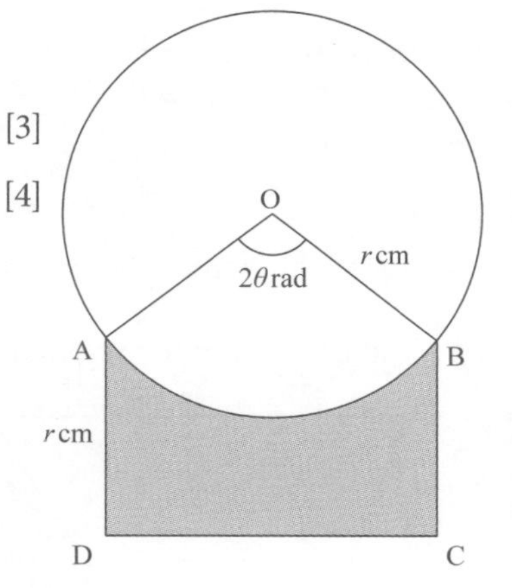

Cambridge International AS & A Level Mathematics 9709 Paper 12 Q4 June 2017

2 **(i)** Show that $\cos^4 x \equiv 1 - 2\sin^2 x + \sin^4 x$. [1]

(ii) Hence, or otherwise, solve the equation $8\sin^4 x + \cos^4 x = 2\cos^2 x$ for $0° \leqslant x \leqslant 360°$. [5]

Cambridge International AS & A Level Mathematics 9709 Paper 11 Q6 November 2016

3 In the diagram, triangle ABC is right-angled at C and M is the midpoint of BC. It is given that angle ABC = $\frac{1}{3}\pi$ radians and angle BAM = θ radians. Denoting the lengths of BM and MC by x,

(i) find AM in terms of x, [3]

(ii) show that $\theta = \frac{1}{6}\pi - \tan^{-1}\left(\frac{1}{2\sqrt{3}}\right)$. [2]

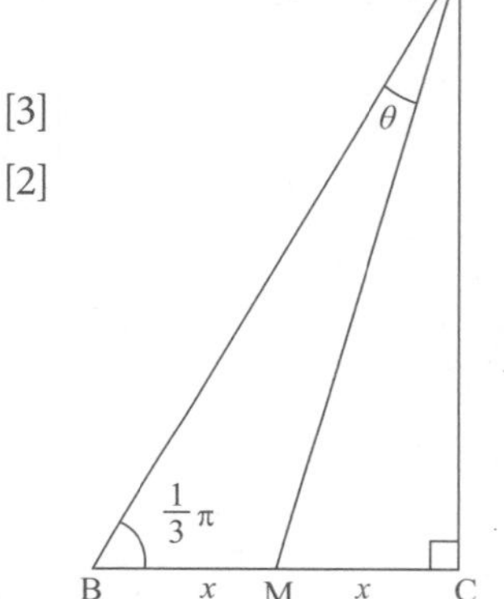

Cambridge International AS & A Level Mathematics 9709 Paper 12 Q5 June 2016

4 In the diagram, OAB is a sector of a circle with centre O and radius r. The point C on OB is such that angle ACO is a right angle. Angle AOB is α radians and is such that AC divides the sector into two regions of equal area.

(i) Show that $\sin\alpha\cos\alpha = \frac{1}{2}\alpha$. [4]

It is given that the solution of the equation in part **(i)** is $\alpha = 0.9477$, correct to 4 decimal places.

(ii) Find the ratio perimeter of region OAC : perimeter of region ACB, giving your answer in the form $k : 1$, where k is given correct to 1 decimal place. [5]

(iii) Find angle AOB in degrees. [1]

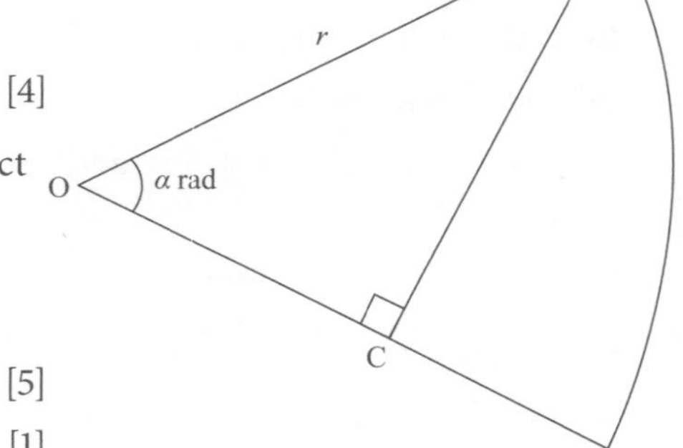

Cambridge International AS & A Level Mathematics 9709 Paper 13 Q11 June 2015

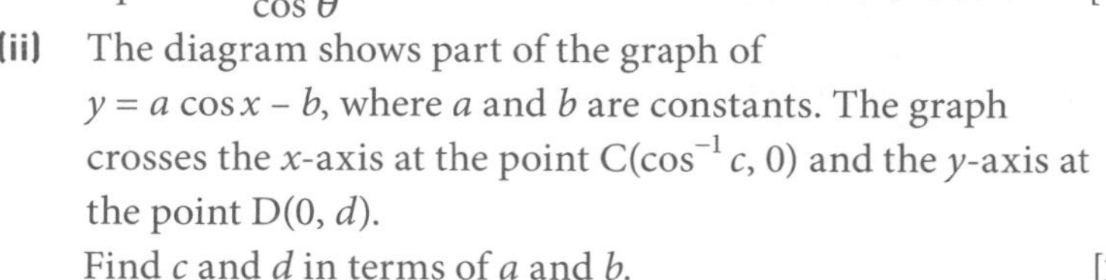

5 (i) Show that the equation $\frac{1}{\cos\theta} + 3\sin\theta\tan\theta + 4 = 0$ can be expressed as $3\cos 2\theta - 4\cos\theta - 4 = 0$, and hence solve the equation $\frac{1}{\cos\theta} + 3\sin\theta\tan\theta + 4 = 0$ for $0° \leqslant \theta \leqslant 360°$. [6]

(ii) The diagram shows part of the graph of $y = a\cos x - b$, where a and b are constants. The graph crosses the x-axis at the point $C(\cos^{-1} c, 0)$ and the y-axis at the point $D(0, d)$.
Find c and d in terms of a and b. [2]

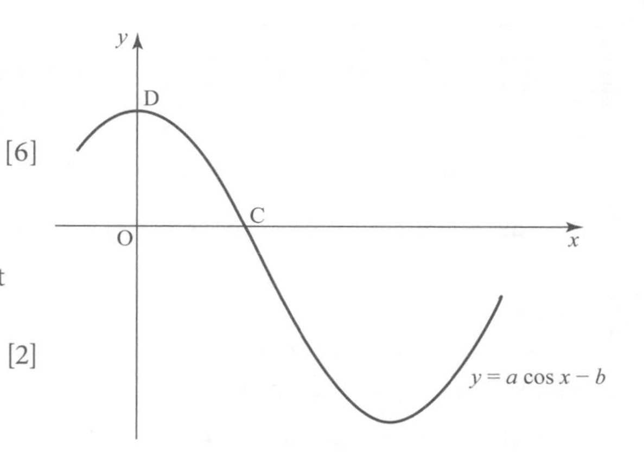

Cambridge International AS & A Level Mathematics 9709 Paper 13 Q7 November 2015

6 The diagram shows a triangle AOB in which OA is 12 cm, OB is 5 cm and angle AOB is a right angle. Point P lies on AB and OP is an arc of a circle with centre A. Point Q lies on AB and OQ is an arc of a circle with centre B.

(i) Show that the angle BAO is 0.3948 radians, correct to 4 decimal places. [1]

(ii) Calculate the area of the shaded region. [5]

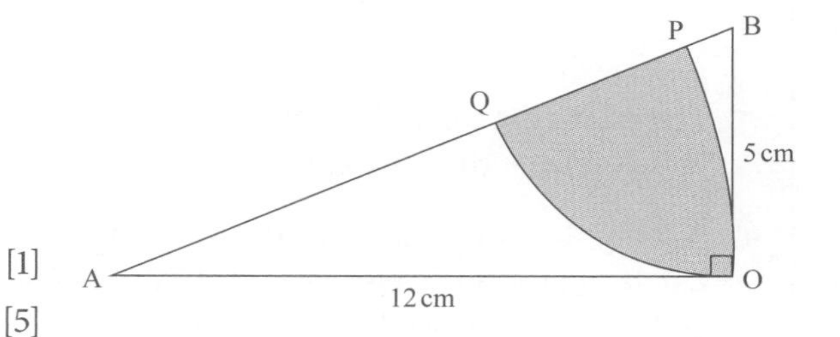

Cambridge International AS & A Level Mathematics 9709 Paper 12 Q2 November 2014

7 (i) Prove the identity $\left(\frac{1}{\sin\theta} - \frac{1}{\tan\theta}\right)^2 \equiv \frac{1-\cos\theta}{1+\cos\theta}$. [3]

(ii) Hence solve the equation $\left(\frac{1}{\sin\theta} - \frac{1}{\tan\theta}\right)^2 = \frac{2}{5}$, for $0° \leqslant \theta \leqslant 360°$. [4]

Cambridge International AS & A Level Mathematics 9709 Paper 13 Q8 June 2011

8 The diagram shows a circle C_1 touching a circle C_2 at a point X. Circle C_1 has centre A and radius 6 cm, and circle C_2 has centre B and radius 10 cm. Points D and E lie on C_1 and C_2 respectively and DE is parallel to AB. Angle DAX $= \frac{1}{3}\pi$ radians and angle EBX $= \theta$ radians.

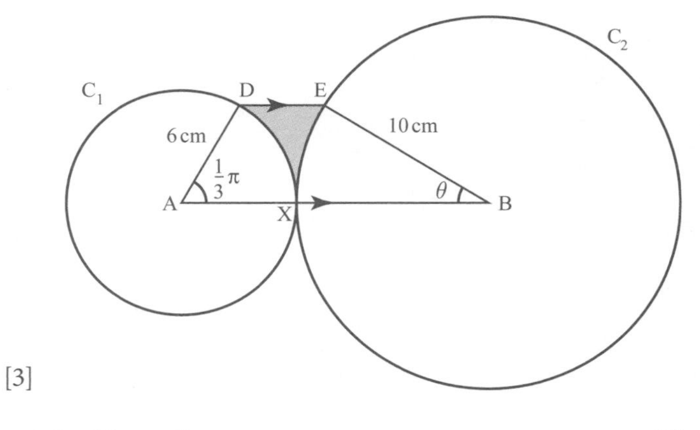

(i) By considering the perpendicular distances of D and E from AB, show that the exact value of θ is $\sin^{-1}\left(\frac{3\sqrt{3}}{10}\right)$. [3]

(ii) Find the perimeter of the shaded region, correct to 4 significant figures. [5]

Cambridge International AS & A Level Mathematics 9709 Paper 12 Q6 November 2011

STRETCH AND CHALLENGE

1 Solve the equation $3\sin 3\theta + 3 = 2\cos^2 3\theta$ for $-180° \leqslant \theta \leqslant 180°$.

2 The height H of the sea above sea level at a certain jetty is given by the equation

$H = 2\sin\frac{\pi}{2}t + 3$, where t is the time in hours from midnight.

(i) Sketch the graph of $H = 2\sin\frac{\pi}{2}t + 3$ for $0 \leqslant t \leqslant 8$.

(ii) Find the period of H.

(iii) At what time will the first high tide occur?

3 The height of water, h, above a reef is modelled by the equation

$h = 9 - 3\cos\left(\frac{\pi}{8}t\right)$, where t is time in hours after low tide.

A ship which has run aground can only be refloated when the water level is 10.5 m above the reef. The refloating can only happen after a low tide which is 9.30 a.m. the next morning. Between what times would it next be possible to refloat the ship?

4 A solar eclipse occurs when the Moon passes between the Earth and the Sun. Assume that the radii of both the Sun and Moon when seen from the Earth are the same.

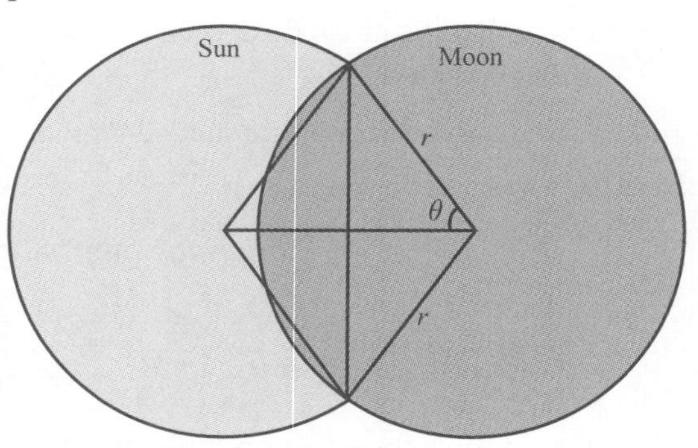

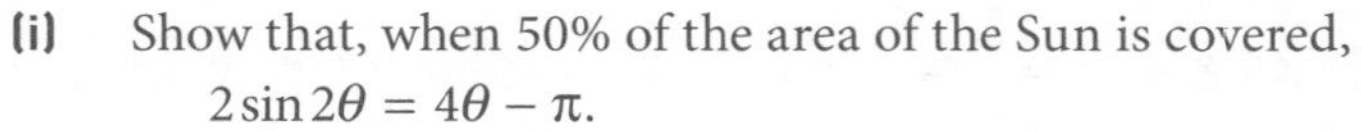

(i) Show that, when 50% of the area of the Sun is covered, $2\sin 2\theta = 4\theta - \pi$.

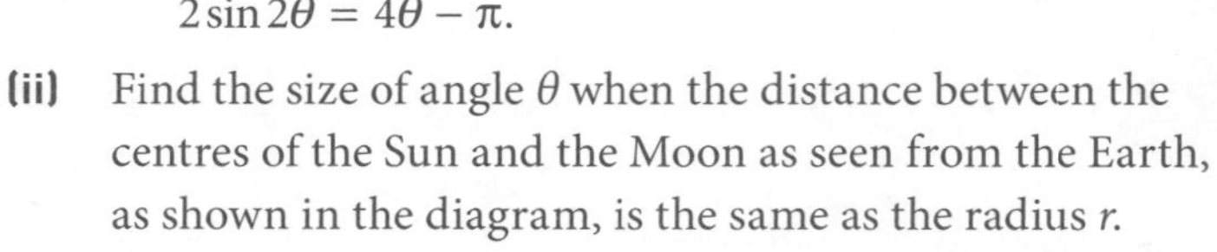

(ii) Find the size of angle θ when the distance between the centres of the Sun and the Moon as seen from the Earth, as shown in the diagram, is the same as the radius r.

(iii) Hence find what percentage area of the Sun is covered when the distance between the centres is the same as the radius r.

Formula sheet

PURE MATHEMATICS

Algebra

For the quadratic equation $ax^2 + bx + c = 0$:

$$x = \frac{-b \pm \sqrt{(b^2 - 4ac)}}{2}$$

For an arithmetic series:

$$u_n = a + (n-1)d, \qquad S_n = \tfrac{1}{2}n(a+l) = \tfrac{1}{2}n\{2a + (n-1)d\}$$

For a geometric series:

$$u_n = ar^{n-1}, \qquad S_n = \frac{a(1-r^n)}{1-r} \quad (r \neq 1), \qquad S_\infty = \frac{a}{1-r} \quad (|r| < 1)$$

Binomial expansion:

$$(a+b)^n = a^n + \binom{n}{1}a^{n-1}b + \binom{n}{2}a^{n-2}b + \binom{n}{3}a^{n-3}b^3 + \ldots + b^n, \text{ where } n \text{ is a positive integer}$$

$$\text{and } \binom{n}{r} = \frac{n!}{r!(n-r)!}$$

Trigonometry

Arc length of circle $= r\theta$ (θ in radians)

Area of sector of circle $= \frac{1}{2}r^2\theta$ (θ in radians)

$$\tan\theta \equiv \frac{\sin\theta}{\cos\theta}$$

$$\cos^2\theta + \sin^2\theta \equiv 1$$

Differentiation

$\mathrm{f}(x)$	$\mathrm{f}'(x)$
x^n	nx^{n-1}

Integration

$\mathrm{f}(x)$	$\int \mathrm{f}(x)\,\mathrm{d}x$
x^n	$\frac{x^{n+1}}{n+1} + c \quad (n \neq -1)$

Reinforce learning and deepen understanding of the key concepts covered in the latest syllabus; an ideal course companion or homework book for use throughout the course.

- Develop and strengthen skills and knowledge with a wealth of additional exercises that perfectly supplement the Student's Book.
- Build confidence with extra practice for each lesson to ensure that a topic is thoroughly understood before moving on.
- Ensure students know what to expect with hundreds of rigorous practice and exam-style questions.
- Keep track of students' work with ready-to-go write-in exercises.
- Save time with all answers available online at: **www.hoddereducation.com/cambridgeextras.**

Use with *Cambridge International AS & A Level Mathematics Pure Mathematics 1 Second edition*

9781510421721

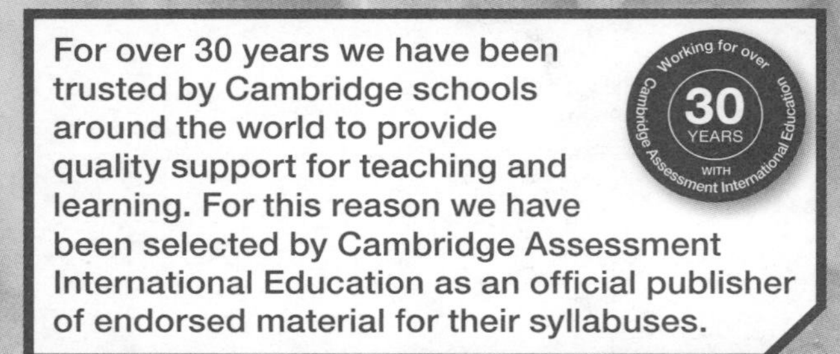

HODDER EDUCATION
e: education@hachette.co.uk
w: hoddereducation.com

ISBN 978-1-5104-2184-4